80 ***The Hardy–Littlewood method***

R. C. VAUGHAN
Professor of Pure Mathematics
Imperial College, University of London

The Hardy–Littlewood method

CAMBRIDGE UNIVERSITY PRESS
CAMBRIDGE
LONDON NEW YORK NEW ROCHELLE
MELBOURNE SYDNEY

Published by the Press Syndicate of the University of Cambridge
The Pitt Building, Trumpington Street, Cambridge CB2 1RP
32 East 57th Street, New York, NY 10022, USA
296 Beaconsfield Parade, Middle Park, Melbourne 3206, Australia

First published 1981

Printed in Great Britain at the
University Press, Cambridge

British Library cataloguing in publication data

Vaughan, R. C.
The Hardy–Littlewood method.—(Cambridge tracts
in mathematics; vol. 80).

1. Numbers, Theory of
I. Title II. Series
512′.73 QA241 80-041206

ISBN 0 521 23439 5

Contents

Preface

There have been two earlier Cambridge Tracts that have touched upon the Hardy–Littlewood method, namely those of Landau, 1937, and Estermann, 1952. However there has been no general account of the method published in the United Kingdom despite the not inconsiderable contribution of English scholars in inventing and developing the method and the numerous monographs that have appeared abroad.

The purpose of this tract is to give an account of the classical forms of the method together with an outline of some of the more recent developments. It has been deemed more desirable to have this particular emphasis as many of the later applications make important use of the classical material.

It would have been useful to devote some space to the work of Davenport on cubic forms, to the joint work of Davenport and Lewis on simultaneous equations, to the work of Rademacher and Siegel that extends the method to algebraic numbers, and to the work of various authors, culminating in the recent work of Schmidt, on bounds for solutions of homogeneous equations and inequalities. However this would have made the tract unwieldy. The interested reader is referred to the Bibliography.

It is assumed that the reader has a familiarity with the elements of number theory, such as is contained in the treatise of Hardy and Wright. Also, in dealing with one or two subjects it is expected that the reader has a working acquaintance with more advanced topics in number theory. Where necessary, reference is given to a standard text on the subject.

The contents of Chapters 2, 3, 4, 5, 9, 10 and 11 have been made the basis of advanced courses offered at Imperial College over a number of years, and could be used as part of any normal postgraduate training in analytic number theory.

Notation

The letter k denotes a natural number, usually with $k \geqslant 2$, and the statements in which ε appear are true for every positive real number ε. The letter p is reserved for prime numbers.

The Vinogradov symbols $\ll$, $\gg$ have their usual meaning, namely that for functions f and g with g taking non-negative real values $f \ll g$ means $|f| \leqslant Cg$ where C is a constant, and if moreover f is also non-negative, then $f \gg g$ means $g \ll f$.

Implicit constants in the O, $\ll$ and $\gg$ notations usually depend on k, s and ε. Additional dependence will be mentioned explicitly.

As usual in number theory, the functions $e(\alpha)$ and $\|\alpha\|$ denote $e^{2\pi i\alpha}$ and $\min_{h \in \mathbb{Z}} |\alpha - h|$ respectively. Occasionally the expression $\min(X, 1/0)$ occurs, and is taken to be X.

The notation $p^r \| n$ is used to mean that p^r is the highest power of p dividing n.

1

Introduction and historical background

1.1 Waring's problem

In 1770 E. Waring asserted without proof in his *Meditationes Algebraicae* that every even natural number is a sum of at most nine positive integral cubes, also a sum of at most 19 biquadrates, and so on. By this it is usually assumed that he believed that for every natural number $k \geqslant 2$ there exists a number s such that every natural number is a sum of at most s kth powers of natural numbers, and that the least such s, say $g(k)$, satisfies $g(3) = 9$, $g(4) = 19$.

It was probably known to Diophantus, albeit in a different form, that every natural number is the sum of at most four squares. The four square theorem was first stated explicitly by Bachet in 1621, and a proof was claimed by Fermat but he died before disclosing it. It was not until 1770 that one was given, by Lagrange, who built on earlier work of Euler. For an account of this theorem see Chapter 20 of Hardy & Wright (1979).

In the 19th century the existence of $g(k)$ was established for many values of k, but it was not until the present century that substantial progress was made. First of all Hilbert (1909*a*, *b*) demonstrated the existence of $g(k)$ for every k by a difficult combinatorial argument based on algebraic identities (see Rieger, 1953*a*, *b*, *c*; Ellison, 1971). His method gives a very poor bound for $g(k)$.

In the early 1920s Hardy and Littlewood introduced an analytic method which has been the basis for numerical work by Dickson, Pillai and others, and has led to an almost complete evaluation of $g(k)$. Since the integer

$$n = 2^k\left[\left(\frac{3}{2}\right)^k\right] - 1$$

is smaller than 3^k it can only be a sum of kth powers of 1 and 2. Clearly the most economical representation is by $[(\frac{3}{2})^k] - 1$ kth powers of 2 and $2^k - 1$ kth powers of 1. Thus

$$g(k) \geqslant 2^k + \left[\left(\frac{3}{2}\right)^k\right] - 2. \tag{1.1}$$

It is very plausible that this always holds with equality, and the current state of knowledge is as follows.

Suppose that $k \neq 4$. It has been shown that when

$$2^k\left\{\left(\frac{3}{2}\right)^k\right\} + \left[\left(\frac{3}{2}\right)^k\right] \leqslant 2^k \tag{1.2}$$

one has

$$g(k) = 2^k + \left[\left(\frac{3}{2}\right)^k\right] - 2 \tag{1.3}$$

but when

$$2^k\left\{\left(\frac{3}{2}\right)^k\right\} + \left[\left(\frac{3}{2}\right)^k\right] > 2^k$$

one has either

$$g(k) = 2^k + \left[\left(\frac{3}{2}\right)^k\right] + \left[\left(\frac{4}{3}\right)^k\right] - 2$$

or

$$g(k) = 2^k + \left[\left(\frac{3}{2}\right)^k\right] + \left[\left(\frac{4}{3}\right)^k\right] - 3$$

according as

$$\left[\left(\frac{4}{3}\right)^k\right]\left[\left(\frac{3}{2}\right)^k\right] + \left[\left(\frac{4}{3}\right)^k\right] + \left[\left(\frac{3}{2}\right)^k\right]$$

is equal to 2^k or is larger than 2^k. For the various contributions to the proof of this, see the Bibliography.

Stemmler (1964) has verified on a computer that (1.2) (and so (1.3)) holds whenever $k \leqslant 200\,000$, and Mahler (1957) has shown that if there are any values of k for which (1.2) is false, then there can only be a finite number of such values. No exceptions are known, and unfortunately the method will not give a bound beyond which there are no exceptions.

Thomas (1974) has shown that $g(4) \leqslant 22$ (whence, by (1.1), $g(4) = 19$,

20, 21 or 22), and Balasubramanian has recently announced $g(4) \leqslant 21$. Thomas has also shown that n is a sum of at most 19 biquadrates whenever $n < 10^{310}$ or $n > 10^{1409}$.

1.2 The Hardy–Littlewood method

Nearly all the above conclusions have been obtained in the following way. A theoretical argument based on the analytic method of Hardy and Littlewood produces a number C_k such that every natural number larger than C_k is the sum of at most s_k kth powers of natural numbers where s_k does not exceed the expected value of $g(k)$. Then a rather tedious, but often very ingenious, calculation enables a check to be made on all the natural numbers not exceeding C_k.

One of the features of the Hardy–Littlewood method is that it can be adapted to attack many other problems of an additive nature. The method has its genesis in a paper of Hardy & Ramanujan (1918) concerned mainly with the partition function, but also dealing with the representation of numbers as sums of squares.

Let $\mathscr{A} = (a_m)$ denote a strictly increasing sequence of non-negative integers and consider

$$F(z) = \sum_{m=1}^{\infty} z^{a_m} \qquad (|z| < 1)$$

and its sth power

$$F(z)^s = \sum_{m_1=1}^{\infty} \dots \sum_{m_s=1}^{\infty} z^{a_{m_1} + \dots + a_{m_s}} = \sum_{n=0}^{\infty} R_s(n) z^n,$$

where $R_s(n)$ is the number of representations of n as the sum of s members of $\mathscr{A}$. The objective is an estimate for $R_s(n)$, at least when n is large. By Cauchy's integral formula

$$R_s(n) = \frac{1}{2\pi i} \int_{\mathscr{C}} F(z)^s z^{-n-1} dz$$

where $\mathscr{C}$ is a circle centre 0 of radius ρ, $0 < \rho < 1$.

Hardy and Ramanujan discovered an alternative way of evaluating the integral when $a_m = m^2$. Suppose that $\rho = 1 - \frac{1}{n}$ and that n is large, and write $e(\alpha) = e^{2\pi i \alpha}$. Then the function F has 'peaks' when $z = \rho e(\alpha)$ is 'close' to the point $e(a/q)$ with q 'not too large'. In fact, F has an asymptotic expansion in the neighbourhood of such points, roughly speaking valid when $|\alpha - a/q| \leqslant 1/(q\sqrt{n})$ and $q \leqslant \sqrt{n}$. By Dirichlet's theorem on diophantine approximation every z under consideration is in some such neighbourhood.

The asymptotic expansion takes the form

$$F\left(\rho e\left(\frac{a}{q}+\beta\right)\right) \sim \frac{C}{q} S(q, a)(1-\rho e(\beta))^{-1/2} \tag{1.4}$$

where

$$S(q, a) = \sum_{m=1}^{q} e(am^2/q).$$

This can be seen by dealing first with the case $\beta = 0$ by partitioning the squares into residue classes modulo q and then applying partial summation. Thus, for $s \geqslant 5$ one can obtain

$$R_s(n) \sim \mathfrak{S}_s(n) J_s(n) \tag{1.5}$$

where

$$\mathfrak{S}_s(n) = \sum_{q=1}^{\infty} \sum_{\substack{a=1 \\ (a,q)=1}}^{q} q^{-s} S(q, a)^s e(-an/q)$$

and

$$J_s(n) = C^s \int_{-1/2}^{1/2} (1-\rho e(\beta))^{-s/2} \rho^{-n} e(-\beta n) d\beta.$$

The integral in $J_s(n)$ is quite easy to estimate, and the series $\mathfrak{S}_s(n)$ reflects certain interesting number theoretic properties of the sequence of squares.

The expansion (1.4) corresponds to a singularity of the series F at $e(a/q)$ on its circle of convergence, and in view of this Hardy and Littlewood coined the terms *singular series* and *singular integral* for $\mathfrak{S}_s(n)$ and $J_s(n)$ respectively.

After the First World War, Hardy & Littlewood (1920, 1921) turned their attention to Waring's problem. Unfortunately, when $a_m = m^k$ with $k \geqslant 3$, they could only show that the expansion corresponding to (1.4) holds when

$$q \leqslant n^{1/k-\varepsilon} \quad \text{and} \quad \left|\alpha - \frac{a}{q}\right| \leqslant q^{-1} n^{1/k-\varepsilon-1},$$

and this only accounts for a small proportion of the points z on $\mathscr{C}$. Since $q^{-1}S(q, a) \to 0$ as $q \to \infty$ (for $(a, q) = 1$) one might hope that at any rate F is small compared with the trivial estimate $(1-\rho)^{-1/k}=n^{1/k}$ on the remaining z, a hope reinforced by the fact that (αm^k) is uniformly distributed modulo 1 when α is irrational. Indeed, Hardy and Littlewood were able to show that F is appreciably smaller than $n^{1/k}$ on the remainder of $\mathscr{C}$ by an alternative argument having its origins in Weyl's (1916) fundamental work on the uniform distribution of sequences, the consequent statement about the size of F often being called Weyl's inequality. They further introduced the terms *major arcs* and *minor arcs* to describe the parts of $\mathscr{C}$ where they used the analogue of (1.4) and Weyl's inequality respectively.

Later Vinogradov (1928*a*) introduced a number of notable refinements, one of which was to replace $F(z)$ by the finite sum

$$f(\alpha)=\sum_{m=1}^{N} e(\alpha m^k) \tag{1.6}$$

where

$$N=[n^{1/k}]. \tag{1.7}$$

Now

$$f(\alpha)^s=\sum_{m=1}^{sn} R_s(m,n)e(\alpha m)$$

where $R_s(m,n)$ is the number of representations of m as the sum of s kth powers, none of which exceed n. Thus

$$R_s(m,n)=R_s(m) \quad (m\leqslant n).$$

Then a special case of Cauchy's integral formula, namely the trivial orthogonality relation

$$\int_0^1 e(\alpha h)\mathrm{d}\alpha=\begin{cases}1 & \text{when } h=0\\ 0 & \text{when } h\neq 0\end{cases} \tag{1.8}$$

gives

$$\int_0^1 f(\alpha)^s e(-\alpha n)\mathrm{d}\alpha=R_s(n). \tag{1.9}$$

It is clear from the discussions above that $g(k)$ is determined by the peculiar demands of a few relatively small exceptional natural numbers. Thus the more interesting problem is that of the estimation of the number $G(k)$, defined for $k\geqslant 2$ to be the least s such that every sufficiently large natural number is the sum of at most s kth powers of natural numbers. It transpires that $G(k)$ is much smaller than $g(k)$ when k is large and this naturally makes its evaluation much more difficult. In fact the value of $G(k)$ is only known when $k=2$ or 4, namely

$$G(2)=4, G(4)=16,$$

the latter result being due to Davenport (1939*c*). Linnik (1943*a*) has shown that $G(3)\leqslant 7$ and Watson (1951) has given an extremely elegant proof of this. When $k>3$ all the best estimates available at present for $G(k)$ have been obtained via the Hardy–Littlewood method. Chapters 2, 4, 5, 6, 7 are devoted to the study of $G(k)$.

1.3 Goldbach's problem

In two letters to Euler in 1742, Goldbach conjectured that every even number is a sum of two primes and every number greater than 2 is a sum of three primes. He included 1 as a prime number, and so in modern times Goldbach's conjectures have become the assertions that every even number greater than 2 is a sum of two primes and every odd number greater than 5 is a sum of three primes.

Hardy & Littlewood (1923*a*,*b*) discovered that their method could also be applied with success to these problems, provided that they assumed the generalized Riemann hypothesis. Thus they were able to show conditionally that every large odd number is a sum of three primes and that almost every even number is a sum of two primes.

In 1937, Vinogradov was able to remove the dependence on the generalized Riemann hypothesis, thereby giving unconditional proofs of the above conclusions. This line of attack on Goldbach's problems is investigated in Chapter 3. However, the nature of the primes, and in particular the problem of their distribution in arithmetic progressions, means that the further refinements of the method (see Montgomery & Vaughan, 1975) are better viewed in the context of multiplicative number theory and have therefore been omitted from this tract.

For many generalizations of the methods described in Chapter 3 see Hua's (1965) monograph.

1.4 Other problems

The last thirty years have seen a large expansion and diversity of the applications of the method, and in Chapters 8, 9, 10, 11 a number of topics have been chosen to illustrate this development. The applications described there, particularly in Chapters 9 and 11 to general forms and inequalities respectively, cover only a small part of the work which has been undertaken in these areas, and should be viewed as an introduction to the original papers listed in the Bibliography.

1.5 Exercises

1 Show that the number $\rho(n)$ of solutions of the equation

$$x_1 + \ldots + x_s = n$$

in non-negative integers $x_1, \ldots, x_s$ is $(-1)^n\binom{-s}{n}$.

2 Show that the sum of the divisors of n, $\sigma(n) = \sum_{m|n} m$, satisfies

$$\sigma(n) = \frac{\pi^2}{6} n \sum_{q=1}^{\infty} q^{-2} c_q(n)$$

where $c_q(n)$ is Ramanujan's sum, i.e.

$$c_q(n) = \sum_{\substack{a=1 \\ (a,q)=1}}^{q} e(an/q).$$

3 Let P,Q denote real numbers with $P > 1$, $Q \geqslant 2P$. Show that the intervals

$$\{\alpha : |\alpha - a/q| \leqslant q^{-1}Q^{-1}\}$$

with $q \leqslant P$ and $(a,q) = 1$ are pairwise disjoint.

2

The simplest upper bound for G(k)

2.1 The definition of major and minor arcs

The introduction of various refinements over the years, most notably by Hua (1938*b*) has led to a simple proof that $G(k) \leqslant 2^k + 1$ which nevertheless illustrates many of the salient features of the Hardy–Littlewood method.

There is a good deal of latitude in the definition of major and minor arcs, and the choice made here is fairly arbitrary.

Let n be large, suppose that N is given by (1.7) and that

$$\nu = \frac{1}{100}, \quad P = N^{\nu}, \tag{2.1}$$

and let δ denote a sufficiently small positive number depending only on k. When $1 \leqslant a \leqslant q \leqslant P$ and $(a,q) = 1$, let

$$\mathfrak{M}(q, a) = \{\alpha : |\alpha - a/q| \leqslant N^{\nu - k}\}. \tag{2.2}$$

The $\mathfrak{M}(q, a)$ are called, for the historical reasons outlined above, the *major arcs*, although in fact they are intervals. Let $\mathfrak{M}$ denote the union of the $\mathfrak{M}(q, a)$. It is convenient to work on the unit interval

$$\mathscr{U} = (N^{\nu - k}, 1 + N^{\nu - k}] \tag{2.3}$$

rather than $(0, 1]$. This avoids any difficulties associated with having only 'half major arcs' at 0 and 1. Observe that $\mathfrak{M} \subset \mathscr{U}$. The set $\mathfrak{m} = \mathscr{U} \setminus \mathfrak{M}$ forms the *minor arcs*.

When $a/q \neq a'/q'$ and $q, q' \leqslant N^{\nu}$, one has

$$\left|\frac{a}{q} - \frac{a'}{q'}\right| \geqslant \frac{1}{qq'} > \left(\frac{1}{q} + \frac{1}{q'}\right) N^{\nu - k}.$$

Thus the $\mathfrak{M}(q, a)$ are pairwise disjoint.

By (1.9) (for brevity the suffix s is dropped)

$$R(n) = \int_{\mathfrak{M}} f(\alpha)^s e(-\alpha n) \mathrm{d}\alpha + \int_{\mathfrak{m}} f(\alpha)^s e(-\alpha n) \mathrm{d}\alpha \tag{2.4}$$

where $f(\alpha)$ is given by (1.6). Before proceeding with the estimation of these integrals it is necessary to establish some auxiliary lemmas.

2.2 Auxiliary lemmas

The method for treating $f(\alpha)$ when $\alpha \in \mathfrak{m}$ can be outlined as follows. When $k = 1$,

$$f(\alpha) = \sum_{m=1}^{N} e(\alpha m^k)$$

is trivial to estimate. In the general case, an argument based on the use of the forward difference operator enables $f(\alpha)$ to be estimated in terms of sums in which m^k is replaced by a polynomial of degree $k - 1$. Then successive applications of this argument reduce the degree to 1.

Lemma 2.1 (Dirichlet) *Let α denote a real number. Then for each real number $X \geqslant 1$ there exists a rational number a/q with $(a, q) = 1$, $1 \leqslant q \leqslant X$ and*

$$|\alpha - a/q| \leqslant 1/(qX).$$

Proof It suffices to prove the result without the condition $(a, q) = 1$.

Let $m = [X]$. The m numbers $\beta_q = \alpha q - [\alpha q]$ $(q = 1, 2, \dots, m)$ all lie in $[0, 1)$. Consider the $m + 1$ intervals

$$B_r = \left[\frac{r-1}{m+1}, \frac{r}{m+1}\right) \qquad (r = 1, 2, \dots, m + 1).$$

If there is a β_q in B_1 or B_{m+1}, then the proof is finished. If not, then one of the $m - 1$ boxes B_r with $2 \leqslant r \leqslant m$ contains at least two of the β_q, say β_u, β_v with $u < v$. Take $q = v - u$, $a = [\alpha v] - [\alpha u]$.

Lemma 2.2 *Suppose that X, Y, α are real numbers with $X \geqslant 1$, $Y \geqslant 1$, and that $|\alpha - a/q| \leqslant q^{-2}$ with $(a, q) = 1$. Then*

$$\sum_{x \leqslant X} \min(XYx^{-1}, \|\alpha x\|^{-1}) \ll XY\left(\frac{1}{q} + \frac{1}{Y} + \frac{q}{XY}\right)\log(2Xq)$$

where $\|\beta\| = \min_{y \in \mathbb{Z}} |\beta - y|$.

Proof Let

$$S = \sum_{x \leqslant X} \min(XYx^{-1}, \|\alpha x\|^{-1}).$$

Clearly

$$S \leqslant \sum_{0 \leqslant j \leqslant X/q} \sum_{r=1}^{q} \min\left(\frac{XY}{qj+r}, \|\alpha(qj+r)\|^{-1}\right).$$

For each j let $y_j = [\alpha j q^2]$, and write $\theta = q^2\alpha - qa$. Then

$$\alpha(qj+r) = (y_j + ar)/q + \{\alpha j q^2\}/q + \theta r q^{-2}.$$

When $j = 0$ and $r \leqslant \frac{1}{2}q$,

$$\|\alpha(qj+r)\| \geqslant \|ar/q\| - 1/(2q) \geqslant \tfrac{1}{2}\|ar/q\|.$$

Otherwise, for each j there are at most $O(1)$ values of r for which $\|\alpha(qj+r)\| \geqslant \frac{1}{2}\|(y_j+ar)/q\|$ fails to hold, and moreover $qj + r \gg q(j+1)$. Therefore

$$S \ll \sum_{1 \leqslant r \leqslant q/2} \|ar/q\|^{-1} + \sum_{0 \leqslant j \leqslant X/q} \left(\frac{XY}{q(j+1)} + \sum_{\substack{r=1 \\ q \nmid y_j + ar}}^{q} \|(y_j+ar)/q\|^{-1}\right)$$

$$\ll XYq^{-1} \sum_{0 \leqslant j \leqslant X} \frac{1}{j+1} + (Xq^{-1}+1) \sum_{1 \leqslant h \leqslant q/2} \frac{q}{h},$$

and the lemma follows easily.

Let Δ_j denote the jth iterate of the forward difference operator, so that for any function ϕ of a real variable α

$$\Delta_1(\phi(\alpha);\beta) = \phi(\alpha+\beta) - \phi(\alpha),$$

$$\Delta_{j+1}(\phi(\alpha);\beta_1,\ldots,\beta_{j+1}) = \Delta_1(\Delta_j(\phi(\alpha);\beta_1,\ldots,\beta_j);\beta_{j+1}).$$

Then it is an easy exercise to show that

$$\Delta_j(\alpha^k;\beta_1,\ldots,\beta_j) = \beta_1 \ldots \beta_j p_j(\alpha;\beta_1,\ldots,\beta_j)$$

where p_j is a polynomial in α of degree $k-j$ which has leading coefficient $k!/(k-j)!$.

The following lemma is an intermediate step in the proofs of both Lemmas 2.4 and 2.5 below.

Lemma 2.3 (Weyl) *Let*

$$T(\phi) = \sum_{x=1}^{Q} e(\phi(x))$$

where ϕ is an arbitrary arithmetical function. Then

$$|T(\phi)|^{2^j} \leqslant (2Q)^{2^j - j - 1} \sum_{|h_1| < Q} \dots \sum_{|h_j| < Q} T_j$$

where

$$T_j = \sum_{x \in I_j} e(\Delta_j(\phi(x); h_1, \dots, h_j))$$

and the intervals $I_j = I_j(h_1, \dots, h_j)$ (possibly empty) satisfy

$$I_1(h_1) \subset [1, Q], I_j(h_1, \dots, h_j) \subset I_{j-1}(h_1, \dots, h_{j-1}).$$

Proof By induction on j. For brevity write $\Delta_j(x)$ for $\Delta_j(\phi(x); h_1, \dots, h_j)$. Obviously

$$\begin{aligned}|T(\phi)|^2 &= \sum_{x=1}^{Q} \sum_{h_1 = 1-x}^{Q-x} e(\Delta_1(x)) \\ &= \sum_{h_1 = 1-Q}^{Q-1} \sum_{x \in I_1} e(\Delta_1(x))\end{aligned}$$

where $I_1 = [1, Q] \cap [1 - h_1, Q - h_1]$.

Now if the conclusion of the lemma is assumed for a particular value of j, then by Cauchy's inequality,

$$|T(\phi)|^{2^{j+1}} \leqslant (2Q)^{2^{j+1} - 2j - 2} (2Q)^j \sum_{h_1, \dots, h_j} |T_j|^2$$

and obviously

$$|T_j|^2 = \sum_{|h| < Q} \sum_{x \in I_{j+1}} e(\Delta_j(x + h) - \Delta_j(x))$$

with $I_{j+1} = I_j \cap \{x : x + h \in I_j\}$.

Lemma 2.4 (Weyl's inequality) *Suppose that* $(a, q) = 1$,

$$|\alpha - a/q| \leqslant q^{-2}, \quad \phi(x) = \alpha x^k + \alpha_1 x^{k-1} + \dots + \alpha_{k-1} x + \alpha_k$$

and

$$T(\phi) = \sum_{x=1}^{Q} e(\phi(x)).$$

Then

$$T(\phi) \ll Q^{1+\varepsilon} (q^{-1} + Q^{-1} + qQ^{-k})^{1/K}$$

where $K = 2^{k-1}$.

Proof By Lemma 2.3 with $j = k-1$ (and Exercise 2.1),

$$|T(\phi)|^K \leqslant (2Q)^{K-k} \times \sum_{h_1} \dots \sum_{\substack{h_{k-1} \\ |h_j| \leqslant Q}} \sum_{x \in I_{k-1}} e(h_1 \dots h_{k-1} p_{k-1}(x; h_1, \dots, h_{k-1}))$$

with

$$p_{k-1}(x; h_1, \dots, h_{k-1}) = k!\alpha(x + \tfrac{1}{2}h_1 + \dots + \tfrac{1}{2}h_{k-1}) + (k-1)!\alpha_1.$$

The terms with $h_1 \dots h_{k-1} = 0$ contribute $\ll Q^{k-1}$. Hence

$$|T(\phi)|^K \ll (2Q)^{K-k}\left(Q^{k-1} + Q^{\varepsilon} \sum_{h=1}^{k!Q^{k-1}} \min(Q, \|\alpha h\|^{-1})\right)$$
$$\ll Q^{K-k+\varepsilon}\left(Q^{k-1} + \sum_{h=1}^{k!Q^{k-1}} \min(Q^k h^{-1}, \|\alpha h\|^{-1})\right).$$

By Lemma 2.2, when $q \leqslant Q^k$ this is

$$\ll Q^{K+2\varepsilon}(q^{-1} + Q^{-1} + qQ^{-k}).$$

The proof is completed by observing that the result is trivial when $q > Q^k$.

Lemma 2.5 (Hua's lemma, 1938*b*) *Suppose that* $1 \leqslant j \leqslant k$. *Then*

$$\int_0^1 |f(\alpha)|^{2^j} \mathrm{d}\alpha \ll N^{2^j - j + \varepsilon}. \tag{2.5}$$

Proof By induction on j. The case $j = 1$ is immediate from Parseval's identity.

Now suppose that (2.5) holds and that $1 \leqslant j \leqslant k-1$. By Lemma 2.3 with $\phi(x) = \alpha x^k$,

$$|f(\alpha)|^{2^j} \leqslant (2N)^{2^j - j - 1} \sum_{h_1} \dots \sum_{\substack{h_j \\ |h_i| \leqslant N}} \sum_{x \in I_j} e(\alpha h_1 \dots h_j p_j(x; h_1, \dots, h_j))$$

where $p_j(x; h_1, \dots, h_j)$ is a polynomial in x of degree $k - j$ with integer coefficients. Hence

$$|f(\alpha)|^{2^j} \ll (2N)^{2^j - j - 1} \sum_h c_h e(\alpha h) \tag{2.6}$$

where c_h is the number of solutions of the equation

$$h_1 \dots h_j p_j(x; h_1, \dots, h_j) = h$$

with $|h_i| < N$ and $x \in I_j$. Clearly

$$c_0 \ll N^j, \quad c_h \ll N^\varepsilon \ (h \neq 0).$$

By writing

$$|f(\alpha)|^{2^j} = f(\alpha)^{2^{j-1}} f(-\alpha)^{2^{j-1}}$$

one also obtains

$$|f(\alpha)|^{2^j} = \sum_h b_h e(-\alpha h) \tag{2.7}$$

where b_h is the number of solutions of

$$x_1^k + \dots + x_{2^{j-1}}^k - y_1^k - \dots - y_{2^{j-1}}^k = h$$

with $x_i, y_i \leqslant N$. Thus

$$\sum_h b_h = f(0)^{2^j} = N^{2^j}$$

and, on the inductive hypothesis,

$$b_0 = \int_0^1 |f(\alpha)|^{2^j} d\alpha \ll N^{2^j - j + \varepsilon}.$$

By (2.6), Parseval's identity, and (2.7),

$$\int_0^1 |f(\alpha)|^{2^{j+1}} d\alpha \ll (2N)^{2^j - j - 1} \sum_h c_h b_h.$$

Moreover,

$$\sum_h c_h b_h \ll c_0 b_0 + N^\varepsilon \sum_{h \neq 0} b_h$$
$$\ll N^j N^{2^j - j + \varepsilon} + N^\varepsilon N^{2^j},$$

which gives the desired conclusion.

Lemma 2.6. *Let $c_1, c_2, \dots$ be any sequence of complex numbers and suppose that F has a continuous derivative on $[0, X]$. Then*

$$\sum_{m \leqslant X} c_m F(m) = F(X) \sum_{m \leqslant X} c_m - \int_0^X F'(\gamma) \sum_{m \leqslant \gamma} c_m d\gamma$$

Proof The lemma follows at once by writing $F(m) = F(X) - \int_m^X F'(\gamma) d\gamma$ and interchanging the order of summation and integration.

2.3 The treatment of the minor arcs

Theorem 2.1 *Suppose that* $s > 2^k$. *Then*

$$\int_{\mathfrak{m}} |f(\alpha)|^s d\alpha \ll n^{s/k - 1 - \delta}.$$

Proof An amount $n^{-1-\delta}$ has to be saved on the trivial estimate $n^{s/k}$. Hua's lemma with $j = k$ saves $n^{\varepsilon - 1}$, and Weyl's inequality is used to save the rest.

Obviously

$$\int_{\mathfrak{m}} |f(\alpha)|^s d\alpha \ll \left(\sup_{\alpha \in \mathfrak{m}} |f(\alpha)|\right)^{s-2^k} \int_0^1 |f(\alpha)|^{2^k} d\alpha. \tag{2.8}$$

Consider an arbitrary point α of $\mathfrak{m}$. By Dirichlet's theorem (Lemma 2.1) there exist a, q with $(a, q) = 1$ and $q \leqslant N^{k-\nu}$, and such that $|\alpha - a/q| \leqslant q^{-1}N^{\nu-k}$. Since $\alpha \in \mathfrak{m} \subset (N^{\nu-k}, 1 - N^{\nu-k})$ it follows that $1 \leqslant a \leqslant q$, whence $q > N^{\nu}$ (for otherwise α would be in $\mathfrak{M}$). Therefore, by Weyl's inequality,

$$f(\alpha) \ll N^{1+\varepsilon}(q^{-1} + N^{-1} + qN^{-k})^{1/K} \ll N^{1+\varepsilon-\nu/K}.$$

This, with (1.7), (2.8) and Hua's lemma, gives the theorem.

2.4 The major arcs

The first step is to obtain a suitable approximation to the generating function f on $\mathfrak{M}(q,a)$ by the auxiliary functions

$$v(\beta) = \sum_{m=1}^{n} \frac{1}{k} m^{1/k-1} e(\beta m), \tag{2.9}$$

$$S(q, a) = \sum_{m=1}^{q} e(am^k/q). \tag{2.10}$$

The function v is that obtained from f by replacing the characteristic function of the kth powers by the probability that m is a kth power. The sum $S(q, a)$ is an extra factor that has to be introduced, when α is close to a/q, because the kth powers, in general, are not uniformly distributed modulo q.

Lemma 2.7 *Suppose that* $1 \leqslant a \leqslant q \leqslant N^{\nu}$, $(a, q) = 1$, $\alpha \in \mathfrak{M}(q, a)$. *Then*

$$f(\alpha) = q^{-1}S(q, a)v(\alpha - a/q) + O(N^{2\nu}).$$

Proof For $Y \geqslant 0$,

$$\sum_{m \leqslant Y} e(am^k/q) = \sum_{r=1}^{q} e(ar^k/q) \sum_{\substack{m \leqslant Y \\ m \equiv r (\mathrm{mod}\ q)}} 1 = Yq^{-1}S(q,a) + O(q),$$

and

$$\sum_{m \leqslant Y^k} \frac{1}{k} m^{1/k - 1} = \int_1^{Y^k} \frac{1}{k} \alpha^{1/k - 1} \mathrm{d}\alpha + O(1) = Y + O(1). \qquad (2.11)$$

Let

$$c_m = \begin{cases} e(am/q) - q^{-1}S(q,a)\frac{1}{k}m^{1/k-1} & \text{when } m \text{ is a } k\text{th power} \\ -q^{-1}S(q,a)\frac{1}{k}m^{1/k-1} & \text{otherwise,} \end{cases}$$

and take $Y = \gamma^{1/k}$. Then

$$\sum_{m \leqslant \gamma} c_m \ll q \qquad (\gamma \geqslant 0).$$

Hence, by Lemma 2.6 with $F(\gamma) = e(\beta\gamma)$,

$$\sum_{m \leqslant X} c_m e(\beta m) \ll (1 + |\beta| X) q.$$

Taking $X = n$, $\beta = \alpha - a/q$ establishes the lemma.

The function v that occurs in Lemma 2.7 is not the only possible choice. Both

$$v_1(\beta) = \int_0^{n^{1/k}} e(\beta\gamma^k)\mathrm{d}\gamma$$

and

$$v_2(\beta) = \sum_{h=0}^{n} \frac{\Gamma(h + 1/k)}{h!k} e(\beta h)$$

would serve equally well. There are arguments for and against each of v, v_1, v_2. It is easier to investigate the analytic behaviour of v_1 than that of v or v_2, and the use of v_2 would avoid some of the technical complications in the evaluation of $J(n)$ below. However v_2 is somewhat artificial and the evaluation of $J(n)$ when v is replaced by v_1 requires Fourier's inversion formula.

That v and v_1 behave in much the same way when β is fairly small can be inferred from (2.11) and Lemma 2.6 Thus

$$\begin{aligned} v(\beta) &= e(\beta n)n^{1/k} - 2\pi i\beta\int_0^n e(\beta\gamma)\gamma^{1/k}\mathrm{d}\gamma + O(1+n|\beta|) \\ &= \int_0^n e(\beta\gamma)\frac{1}{k}\gamma^{1/k-1}\mathrm{d}\gamma + O(1+n|\beta|) \\ &= v_1(\beta) + O(1+n|\beta|). \end{aligned}$$

Let

$$V(\alpha, q, a) = q^{-1}S(q, a)v(\alpha - a/q). \tag{2.12}$$

Then, by Lemma 2.7, when $\alpha\in\mathfrak{M}(q, a)$,

$$f(\alpha)^s - V(\alpha, q, a)^s \ll N^{s-1}|f(\alpha) - V(\alpha, q, a)| \ll N^{s-1+2\nu}.$$

Therefore

$$\sum_{q\leqslant N^\nu}\sum_{\substack{a=1\\(a,q)=1}}^{q}\int_{\mathfrak{M}(q,a)}|f(\alpha)^s - V(\alpha, q, a)^s|\mathrm{d}\alpha \ll N^{s-k-1+5\nu}$$

Hence there is a positive number δ, depending only on k, such that

$$\int_{\mathfrak{M}} f(\alpha)^s e(-\alpha n)\mathrm{d}\alpha = \mathrm{R}^*(n) + O(n^{s/k-1-\delta}) \tag{2.13}$$

where

$$R^*(n) = \sum_{q\leqslant N^\nu}\sum_{\substack{a=1\\(a,q)=1}}^{q}\int_{\mathfrak{M}(q,a)} V(\alpha, q, a)^s e(-\alpha n)\mathrm{d}\alpha.$$

By (2.2) and (2.12), $R^*(n)$ factorizes as

$$R^*(n) = \mathfrak{S}(n, N^\nu)J^*(n), \tag{2.14}$$

where

$$\mathfrak{S}(n,Q) = \sum_{q\leqslant Q}\sum_{\substack{a=1\\(a,q)=1}}^{q}(q^{-1}S(q, a))^s e(-an/q)$$

and

$$J^*(n) = \int_{-N^{\nu-k}}^{N^{\nu-k}} v(\beta)^s e(-\beta n)\mathrm{d}\beta. \tag{2.15}$$

The series $\mathfrak{S}(n,Q)$ and integral $J^*(n)$ are best evaluated by first completing the series to infinity and then replacing the interval of integration by a unit interval.

Let

$$S(q) = \sum_{\substack{a=1 \\ (a,q)=1}}^{q} (q^{-1}S(q,a))^s e(-an/q). \tag{2.16}$$

By Weyl's inequality, $S(q,a) \ll q^{1+\varepsilon-1/K}$ provided that $(a,q)=1$. Therefore

$$S(q) \ll q^{(\varepsilon-1/K)s+1} \ll q^{-1-2^{-k}} \tag{2.17}$$

when $s \geqslant 2^k + 1$ and ε is sufficiently small. Therefore

$$\mathfrak{S}(n) = \sum_{q=1}^{\infty} S(q) \tag{2.18}$$

converges absolutely, and uniformly with respect to n, and

$$\mathfrak{S}(n, N^\nu) - \mathfrak{S}(n) \ll n^{-\delta}.$$

Hence, by (2.14),

$$R^*(n) = (\mathfrak{S}(n) + O(n^{-\delta}))J^*(n) \text{ and } \mathfrak{S}(n) \ll 1. \tag{2.19}$$

In order to extend the interval of integration in $J^*(n)$, as mentioned above, it is necessary to estimate the rate of decay of $v(\beta)$ as $|\beta|$ increases from 0 to $\frac{1}{2}$.

Lemma 2.8 *Suppose that* $|\beta| \leqslant \frac{1}{2}$. *Then*

$$v(\beta) \ll \min(n^{1/k}, |\beta|^{-1/k}).$$

The same conclusion holds for the functions v_1 and v_2 discussed above, the proofs being similar in each case, and the result for v_1 holding for all real β.

Proof This is by Abel summation. By (2.11),

$$\sum_{r=1}^{m} \frac{1}{k} r^{1/k-1} = m^{1/k} + O(1),$$

and the lemma follows at once when $|\beta| \leqslant 1/n$.

Now suppose that $|\beta| > 1/n$ and let $M = [|\beta|^{-1}]$. Then the terms in

$$v(\beta) = \sum_{m=1}^{n} \frac{1}{k} m^{1/k-1} e(\beta m)$$

with $m \leqslant M$ contribute $\ll M^{1/k} \leqslant |\beta|^{-1/k}$. To estimate the remaining terms, let

$$S_m = \sum_{r=1}^{m} e(\beta r),\ c_m = \frac{1}{k} m^{1/k-1}.$$

Then

$$\sum_{m=M+1}^{n} \frac{1}{k} m^{1/k-1} e(\beta m) = c_{n+1} S_n - c_{M+1} S_M + \sum_{m=M+1}^{n} (c_m - c_{m+1}) S_m.$$

Since $|S_m| \leqslant 1/(2|\beta|)$ and c_m is a decreasing sequence one has

$$\sum_{m=M+1}^{n} \frac{1}{k} m^{1/k-1} e(\beta m) \ll c_{M+1} |\beta|^{-1} < |\beta|^{-1/k}$$

as required.

Let

$$J(n) = \int_{-1/2}^{1/2} v(\beta)^s e(-\beta n) \mathrm{d}\beta. \tag{2.20}$$

Then, by (2.15) and Lemma 2.8,

$$J(n) \ll \int_0^\infty \min(n^{s/k}, \beta^{-s/k}) \mathrm{d}\beta \ll n^{s/k-1}$$

and

$$J^*(n) - J(n) \ll \int_{N^{\nu-k}}^\infty \beta^{-s/k} \mathrm{d}\beta \ll n^{s/k-1-\delta}$$

provided that $s > k$. Hence, by (2.17),

$$R^*(n) = \mathfrak{S}(n) J(n) + O(n^{s/k-1-\delta}). \tag{2.21}$$

This coupled with (2.4), Theorem 2.1 and (2.13) gives

Theorem 2.2 *Suppose that* $s > 2^k$. *Then*

$$R(n) = \mathfrak{S}(n) J(n) + O(n^{s/k-1-\delta}).$$

2.5 The singular integral

The singular integral is estimated by induction on s. The following lemma has the dual rôle of sparking off the inductive process and providing the inductive step.

Lemma 2.9 *Suppose that α, β are real numbers with $\alpha \geqslant \beta > 0$, $\beta \leqslant 1$. Then*

$$\sum_{m=1}^{n-1} m^{\beta-1}(n-m)^{\alpha-1} = n^{\beta+\alpha-1}\left(\frac{\Gamma(\beta)\Gamma(\alpha)}{\Gamma(\beta+\alpha)} + O(n^{-\beta})\right),$$

where the implicit constant depends at most on α and β.

Proof Consider the function

$$\phi(\gamma) = \gamma^{\beta-1}(n-\gamma)^{\alpha-1}.$$

On $(0,n)$, ϕ has at most one stationary point. Thus $(0,n)$ can be divided into two intervals $(0,X)$, (X,n) (one of which may be empty) such that ϕ is increasing on one and decreasing on the other. Therefore

$$\sum_{m=1}^{n-1} \phi(m) = \int_0^n \phi(\gamma)\mathrm{d}\gamma + O(n^{\alpha-1} + n^{\beta+\alpha-2})$$

$$= \frac{\Gamma(\beta)\Gamma(\alpha)}{\Gamma(\beta+\alpha)} n^{\beta+\alpha-1} + O(n^{\alpha-1})$$

Theorem 2.3 *Suppose that $s \geqslant 2$. Then*

$$J(n) = \Gamma\left(1+\frac{1}{k}\right)^s \Gamma\left(\frac{s}{k}\right)^{-1} n^{s/k-1}(1 + O(n^{-1/k})). \qquad (2.22)$$

Proof By (2.9) and (2.20),

$$J(n) = J_s(n) = \sum_{\substack{m_1=1 \\ m_1+\ldots+m_s=n}}^{n} \ldots \sum_{m_s=1}^{n} k^{-s}(m_1 \ldots m_s)^{1/k-1}.$$

When $s = 2$, Lemma 2.9 gives the theorem at once. Suppose the theorem holds for some $s \geqslant 2$. Then

$$J_{s+1}(n) = \sum_{m=1}^{n-1} \frac{1}{k} m^{1/k-1} J_s(n-m)$$

$$= \Gamma\left(1+\frac{1}{k}\right)^s \Gamma\left(\frac{s}{k}\right)^{-1} k^{-1} \sum_{m=1}^{n-1} m^{1/k-1}(n-m)^{s/k-1}$$

$$+ O\left(\sum_{m=1}^{n-1} m^{1/k-1}(n-m)^{(s-1)/k-1}\right).$$

Lemma 2.9 now gives the case $s + 1$.

2.6 The singular series

The singular series reflects the distribution of the kth power residues modulo q. Before investigating the properties of $\mathfrak{S}(n)$ it is necessary to examine $S(q, a)$ and $S(q)$.

Lemma 2.10 *Suppose that* $(a, q) = (b, r) = (q, r) = 1$. *Then*

$$S(qr, ar + bq) = S(q, a)S(r, b).$$

Proof By Euclid's algorithm, each residue class m modulo qr can be represented uniquely in the form $tr + uq$ with $1 \leqslant t \leqslant q$ and $1 \leqslant u \leqslant r$. Therefore, by (2.10),

$$S(qr, ar + bq) = \sum_{t=1}^{q} \sum_{u=1}^{r} e(at^k r^k/q + bu^k q^k/r)$$

Moreover tr and uq run over complete residue classes to the moduli q and r respectively. Hence the lemma.

Lemma 2.11 *The function* $S(q)$ *is multiplicative.*

Proof Suppose that $(q, r) = 1$. Then, by (2.16) and Lemma 2.10,

$$S(qr) = \sum_{\substack{a=1 \\ (a,q)=1}}^{q} \sum_{\substack{b=1 \\ (b,r)=1}}^{r} q^{-s} r^{-s} S(qr, ar + bq)^s e(-(ar + bq)n/(qr))$$
$$= S(q)S(r).$$

For each prime p, define formally

$$T(p) = \sum_{h=0}^{\infty} S(p^h). \tag{2.23}$$

Theorem 2.4 *Suppose that* $s > 2^k$. *Then* $T(p)$ *converges absolutely, so does* $\prod_p T(p)$, *and*

$$\mathfrak{S}(n) = \prod_p T(p).$$

Moreover there is a positive number C, *depending only on* k, *such that*

$$\tfrac{1}{2} < \prod_{p \geqslant C} T(p) < \tfrac{3}{2}.$$

Proof This follows easily from (2.17), Lemma 2.11 and the elementary theory of series of multiplicative functions (see Theorem 286 of Hardy & Wright, 1979). Note that, by (2.16) and (2.10), replacing a by $-a$ in the definition of $S(q)$ gives $S(q) = \bar{S}(q)$. Thus $S(q)$, and so $T(p)$, is real.

It remains to treat $T(p)$ when $p \leqslant C$. There is a close connection between T and the number $M_n(q)$ of solutions of the congruence

$$m_1^k + \ldots + m_s^k \equiv n \pmod{q}$$

with $1 \leqslant m_j \leqslant q$.

Lemma 2.12 *For each natural number q,*

$$\sum_{d|q} S(d) = q^{1-s} M_n(q).$$

Observe that if $q = p^l$, then the left-hand side is

$$\sum_{h=0}^{l} S(p^h)$$

and thus, by (2.23),

$$T(p) = \lim_{l \to \infty} p^{l(1-s)} M_n(p^l)$$

whenever either this limit or the limit in (2.23) exists.

Proof The orthogonality relation

$$\frac{1}{q} \sum_{r=1}^{q} e(hr/q) = \begin{cases} 1 & q \mid h, \\ 0 & q \nmid h, \end{cases}$$

implies that

$$M_n(q) = \frac{1}{q} \sum_{r=1}^{q} \sum_{m_1=1}^{q} \ldots \sum_{m_s=1}^{q} e(r(m_1^k + \ldots + m_s^k - n)/q).$$

Now the sum over r is rearranged into subsums according to the value of (r,q). The general term in each subsum is a periodic function of m_j with period $q/(r, q) = d$, say. Hence

$$M_n(q) = \frac{1}{q} \sum_{d|q} \sum_{\substack{a=1 \\ (a,d)=1}}^{d} \left(\frac{q}{d}\right)^s \sum_{m_1=1}^{d} \ldots \sum_{m_s=1}^{d} e(a(m_1^k + \ldots + m_s^k - n)/d)$$

and the lemma follows from (2.10) and (2.16).

Before proceeding further it is useful to summarize some consequences of the theory of the multiplicative structure of the reduced residue classes modulo p^t. For an exposition of this theory, see Chapter 6 of Vinogradov (1954), or Chapter 10 of Apostol (1976).

The number of kth power residues modulo p^t, i.e. residues of the form x^k with $p \nmid x$, is $\phi(p^t)/(k, \phi(p^t))$ when p is odd or $t = 1$ or k is odd, and $2^{t-2}/(k, 2^{t-2})$ when $t \geqslant 2$ and p and k are both even. (Here ϕ denotes Euler's function.) Thus when p divides k to a high power the kth power residues modulo p^t are comparatively scarce, and so $M_n(p^t)$ is relatively difficult to estimate. It is convenient, therefore, to define $\tau = \tau(p)$ to be the highest power of p that divides k,

$$p^\tau \| k \tag{2.24}$$

and to write

$$\gamma = \gamma(p) = \begin{cases} \tau + 1 & \text{when } p > 2 \text{ or when } p = 2 \text{ and } \tau = 0, \\ \tau + 2 & \text{when } p = 2 \text{ and } \tau > 0. \end{cases} \tag{2.25}$$

Thus the number of kth power residues modulo p^γ is $\phi(p^{\tau+1})/(k, \phi(p^{\tau+1}))$, and the number of solutions of the congruence

$$x^k \equiv a \pmod{p^\gamma}$$

when $p \nmid a$ is 0 or $p^{\gamma-\tau-1}(k, \phi(p^{\tau+1}))$. Also, if a is a kth power residue modulo p^γ, then it is a kth power residue modulo p^t for every t.

Let $M_n^*(q)$ denote the number of solutions of the congruence

$$x_1^k + \ldots + x_s^k \equiv n \pmod q \tag{2.26}$$

with $(x_1, q) = 1$.

Lemma 2.13 *Suppose that* $M_n^*(p^\gamma) > 0$ *and* $t \geqslant \gamma$. *Then*

$$M_n(p^t) \geqslant p^{(t-\gamma)(s-1)}.$$

Proof Consider any solution of

$$x_1^k \equiv n - x_2^k - \ldots - x_s^k \pmod{p^\gamma}$$

with $p \nmid x_1$. Then $p^{(t-\gamma)(s-1)}$ solutions of

$$y_1^k \equiv n - y_2^k - \ldots - y_s^k \pmod{p^t}$$

can be constructed by choosing $y_2, \ldots, y_s$ so that $y_j \equiv x_j \pmod{p^\gamma}$.

Then $n - y_2^k - \dots - y_s^k$ is a kth power residue modulo p^γ, and hence also modulo p^t.

The following lemma is useful in establishing the solubility of (2.26).

Lemma 2.14 (Cauchy, 1813; Davenport, 1935; Chowla, 1935*a*). *Let $\mathscr{A}$, $\mathscr{B}$ respectively denote sets of r, s residue classes modulo q. Suppose further that $0 \in \mathscr{B}$ and that for every $b \in \mathscr{B}$ with $b \not\equiv 0 \pmod q$ one has $(b, q) = 1$. Let $\mathscr{A} + \mathscr{B}$ denote the set of residue classes modulo q of the form $a + b$ with $a \in \mathscr{A}$ and $b \in \mathscr{B}$. Then*

$$\operatorname{card}(\mathscr{A} + \mathscr{B}) \geqslant \min(q, r + s - 1).$$

Proof It can be supposed that $r + s - 1 \leqslant q$, for otherwise one can simply remove $s - (q - r + 1)$ elements from $\mathscr{B}$. The case $r = q$ is trivial so it may be assumed further that $r < q$. The proof now proceeds by induction on s. The case $s = 1$ is trivial. Suppose that $s > 1$ and that the conclusion holds whenever $\operatorname{card} \mathscr{B} < s$. Now there exist $c \in \mathscr{A}, b \in \mathscr{B}$ such that $c + b \notin \mathscr{A}$, for otherwise for each $b \in \mathscr{B}, a + b$ would range over $\mathscr{A}$ as a does, in which case

$$\sum_{a \in \mathscr{A}} (a + b) \equiv \sum_{a \in \mathscr{A}} a \pmod q, \quad rb \equiv 0 \pmod q.$$

Let $\mathscr{C} = \{b : b \in \mathscr{B}, c + b \notin \mathscr{A}\}$, $\mathscr{A}_1 = \mathscr{A} \cup (\{c\} + \mathscr{C})$, $\mathscr{B}_1 = \mathscr{B} \setminus \mathscr{C}$. Then $1 \leqslant \operatorname{card} \mathscr{B}_1 < s$, $\operatorname{card} \mathscr{A}_1 + \operatorname{card} \mathscr{B}_1 = r + s$, and

$$\mathscr{A}_1 + \mathscr{B}_1 = (\mathscr{A} + \mathscr{B}_1) \cup ((\{c\} + \mathscr{B}_1) + \mathscr{C}) \subset \mathscr{A} + \mathscr{B}.$$

Lemma 2.15 *Suppose that $s \geqslant \dfrac{p}{p-1}(k, p^\tau(p-1))$ when $\gamma = \tau + 1$, that $s \geqslant 2^{\tau+2}$ when $\gamma = \tau + 2$ and $k > 2$, and that $s \geqslant 5$ when $p = k = 2$. Then $M_n^*(p^\gamma) > 0$ for every n.*

Proof When $\gamma = \tau + 1$ the lemma follows by repeated application of Lemma 2.14. When $p = 2$ the result is trivial, for when $k > 2$ one has $s \geqslant 2^\gamma$ and the congruence can be satisfied by taking the x_j to be 0 or 1, and when $k = 2$ the congruence $x_1^k + \dots + x_5^k \equiv n \pmod 8$ is easily seen to be soluble with $2 \nmid x_1$.

Collecting together the conclusions of Theorem 2.3 and Lemmas 2.12, 2.13 and 2.15 gives

Theorem 2.5 *Suppose that $s > 2^k$. Then $\mathfrak{S}(n) \gg 1$.*

2.7 Summary

By (2.19) and Theorems 2.2, 2.3 and 2.5 one has

Theorem 2.6 *When $s > 2^k$ the number of representations, $R(n)$, of n as a sum of s kth powers of natural numbers satisfies*

$$R(n) = \Gamma\left(1 + \frac{1}{k}\right)^s \Gamma\left(\frac{s}{k}\right)^{-1} n^{s/k - 1}\mathfrak{S}(n) + O(n^{s/k - 1 - \delta}) \quad (2.27)$$

where $\mathfrak{S}(n) \gg 1$.

Corollary $G(k) \leqslant 2^k + 1$.

The asymptotic formula (2.27) probably holds whenever $s \geqslant k + 1$. The bound $s > 2^k$ has been lowered when $k > 10$. For this see Chapter 5. However, when $3 \leqslant k \leqslant 10$ no improvement is known. Indeed, it would be a major breakthrough to obtain (2.27) when $k = 3$ and $s = 8$. This would follow if it could be shown that

$$\int_0^1 \left| \sum_{x=1}^{N} e(\alpha x^3) \right|^6 d\alpha \ll N^{7/2 - \delta}. \quad (2.28)$$

There is a conjecture that

$$\int_0^1 |f(\alpha)|^{2s} d\alpha \ll N^{\varepsilon} \min(N^s, N^{2s - k}) \quad (2.29)$$

and this would imply that (2.27) holds whenever $s \geqslant 2k + 1$.

Let $k > 2$. Hardy & Littlewood (1922) define $\Gamma(k)$ to be the least s such that for every prime p there is a positive number $C(p)$ such that $T(p) \geqslant C(p)$ uniformly in n. In a later paper, Hardy & Littlewood (1925), they show that $\mathfrak{S}(n) \gg 1$ whenever $s \geqslant \max(\Gamma(k), 4)$.

If one defines $\Gamma(k)$ to be the least s such that for every q and n the congruence

$$x_1^k + \ldots + x_s^k \equiv n \pmod{q}$$

is soluble with $(x_1, q) = 1$, then the proof of their Theorem 1, Hardy & Littlewood (1928), shows that $\Gamma_0(k) = \Gamma(k)$. They conjecture that $\Gamma(k) \to \infty$ as $k \to \infty$, but it is still not even known whether $\liminf_{k \to \infty} \Gamma(k) \geqslant 4$.

2.8 Exercises

1 Show that for $1 \leqslant j \leqslant k$ the jth iterate Δ_j of the forward difference operator satisfies

$$\Delta_j(\alpha^k; \beta_1, \ldots, \beta_j) = \sum_{\substack{l_0, l_1, \ldots, l_j \\ l_0 \geqslant 0, l_1 \geqslant 1, \ldots, l_j \geqslant 1 \\ l_0 + l_1 + \ldots + l_j = k}} \frac{k!}{l_0! l_1! \ldots l_j!} \alpha^{l_0} \beta_1^{l_1} \ldots \beta_j^{l_j}$$

$$= \beta_1 \ldots \beta_k p_j(\alpha; \beta_1, \ldots, \beta_j)$$

where p_j is a polynomial in α of degree $k - j$ and having leading coefficient $k!/(k-j)!$.

2 Show that, when $k > 2$, $G(k) \geqslant \max(k+1, \Gamma_0(k))$.

3 Show that every large natural number is the sum of one square and seven cubes.

4 Show that for $s \geqslant 2$

$$\int_0^1 |f(\alpha)|^s \mathrm{d}\alpha \gg \max(N^{s-k}, N^{s/2}).$$

5 Show that the number R of solutions of

$$x_1^2 + y_1^4 + y_2^4 = x_2^2 + y_3^4 + y_4^4$$

with $x_i \leqslant n^{1/2}$, $y_i \leqslant n^{1/4}$ satisfies $R \ll n^{1+\varepsilon}$. Obtain an asymptotic formula for the number of representations of a number as the sum of two squares, four fourth powers and a kth power.

6 Let

$$v_1(\beta) = \int_0^{n^{1/k}} e(\beta \gamma^k) \mathrm{d}\gamma, \quad v_2(\beta) = \sum_{h=0}^{n} \frac{\Gamma(h + 1/k)}{h! k} e(\beta h).$$

Show that

$$\int_{-\infty}^{\infty} v_1(\beta)^s e(-\beta n) \mathrm{d}\beta \quad \text{and} \quad \int_0^1 v_2(\beta)^s e(-\beta n) \mathrm{d}\beta$$

both equal $\Gamma\left(1 + \frac{1}{k}\right)^s \Gamma\left(\frac{s}{k}\right)^{-1} n^{s/k - 1}$ asymptotically as $n \to \infty$.

3

Goldbach's problems

3.1 The ternary Goldbach problem

Vinogradov's attack on Goldbach's ternary problem follows the pattern of the previous chapter, but this time with

$$f(\alpha) = \sum_{p \leqslant n} (\log p) e(\alpha p). \tag{3.1}$$

The poor current state of knowledge concerning the distribution of primes in arithmetic progressions demands that the major arcs be rather sparse. The principal difficulty then lies on the minor arcs and the establishment of a suitable analogue of Weyl's inequality.

Let B denote a positive constant, and for n sufficiently large write

$$P = (\log n)^B. \tag{3.2}$$

When $1 \leqslant a \leqslant q \leqslant P$ and $(a, q) = 1$, let

$$\mathfrak{M}(q, a) = \{\alpha : |\alpha - a/q| \leqslant Pn^{-1}\} \tag{3.3}$$

denote a typical major arc and write $\mathfrak{M}$ for their union. Since n is large, the major arcs are disjoint and lie in

$$\mathscr{U} = (Pn^{-1}, 1 + Pn^{-1}].$$

Let $\mathfrak{m} = \mathscr{U} \setminus \mathfrak{M}$. Then, by (3.1),

$$R(n) = \int_{\mathscr{U}} f(\alpha)^3 e(-n\alpha)\mathrm{d}\alpha$$
$$= \int_{\mathfrak{M}} f(\alpha)^3 e(-n\alpha)\mathrm{d}\alpha + \int_{\mathfrak{m}} f(\alpha)^3 e(-n\alpha)\mathrm{d}\alpha \tag{3.4}$$

where

$$R(n) = \sum_{\substack{p_1, p_2, p_3 \\ p_1 + p_2 + p_3 = n}} (\log p_1)(\log p_2)(\log p_3). \tag{3.5}$$

The treatment of the minor arcs rests principally on the following theorem

Theorem 3.1 *Suppose that* $(a, q) = 1$, $q \leqslant n$ *and* $|\alpha - a/q| \leqslant q^{-2}$. *Then*

$$f(\alpha) \ll (\log n)^4 (nq^{-1/2} + n^{4/5} + n^{1/2}q^{1/2}).$$

Proof Let

$$\tau_x = \sum_{\substack{d|x \\ d \leqslant X}} \mu(d)$$

where μ is Möbius's function. Then taking $X = n^{2/5}$ and $\lambda(x, y) = \Lambda(y)e(\alpha xy)$ in the identity

$$\sum_{X < y \leqslant n} \lambda(1, y) + \sum_{X < x \leqslant n} \sum_{X < y \leqslant n/x} \tau_x \lambda(x, y)$$

$$= \sum_{d \leqslant X} \sum_{X < y \leqslant n/d} \sum_{z \leqslant n/(yd)} \mu(d)\lambda(dz, y)$$

gives

$$f(\alpha) = S_1 - S_2 - S_3 + O(n^{1/2}),$$

where

$$S_1 = \sum_{x \leqslant X} \sum_{y \leqslant n/x} \mu(x)(\log y)\, e(\alpha xy),$$

$$S_2 = \sum_{x \leqslant X^2} \sum_{y \leqslant n/x} c_x e(\alpha xy) \quad \text{with } c_x = \sum_{d \leqslant X} \sum_{\substack{y \leqslant X \\ dy = x}} \mu(d)\Lambda(y),$$

$$S_3 = \sum_{x > X} \sum_{\substack{y > X \\ xy \leqslant n}} \tau_x \Lambda(y) e(\alpha xy).$$

Here Λ is von Mangoldt's function, and the identity follows by observing that $\tau_x = 0$ $(1 < x \leqslant X)$ and inverting the order of summation.

The inner sum in S_1 is

$$\mu(x) \int_1^{n/x} \sum_{\gamma < y \leqslant n/x} e(\alpha xy) \frac{d\gamma}{\gamma},$$

and $c_x \ll \log x$. Hence

$$S_1, S_2 \ll (\log n) \sum_{x \leqslant X^2} \min(n/x, \|\alpha x\|^{-1}).$$

Therefore, by Lemma 2.2,

$$S_1, S_2 \ll (\log n)^2 (nq^{-1} + n^{4/5} + q).$$

Thus it remains to estimate S_3.

Let $\mathscr{A} = \{X, 2X, 4X, \ldots, 2^k X : 2^k X^2 < n \leqslant 2^{k+1} X^2\}$.

Then

$$S_3 = \sum_{Y \in \mathscr{A}} S(Y)$$

where

$$S(Y) = \sum_{Y < x \leqslant 2Y} \sum_{X < y \leqslant n/x} \tau_x \Lambda(y) e(\alpha x y).$$

By Cauchy's inequality,

$$|S(Y)|^2 \ll \left(\sum_{x \leqslant 2Y} d(x)^2 \right) \sum_{Y < x \leqslant 2Y} \left| \sum_{X < y \leq n/x} \Lambda(y) e(\alpha x y) \right|^2 .$$

It is easily shown that $\sum_{x \leqslant Z} d(x)^2 \ll Z(\log 2Z)^3$. Hence

$$|S(Y)|^2 \ll Y(\log n)^5 \sum_{y \leqslant n/Y} \sum_{z \leqslant n/Y} \min(Y, \|\alpha(y-z)\|^{-1}).$$

Thus, by Lemma 2.2,

$$|S(Y)|^2 \ll n(\log n)^6 (nq^{-1} + Y + n/Y + q)$$

which gives

$$\begin{aligned} S_3 &\ll \sum_{Y \in \mathscr{A}} (\log n)^3 (nq^{-1/2} + n^{1/2} Y^{1/2} + nY^{-1/2} + n^{1/2} q^{1/2}) \\ &\ll (\log n)^4 (nq^{-1/2} + n^{4/5} + n^{1/2} q^{1/2}) \end{aligned}$$

as required.

To estimate

$$\int_{\mathfrak{m}} f(\alpha)^3 e(-\alpha n) d\alpha$$

it is now only necessary to make two observations. First that Parseval's identity and elementary prime number theory together give

$$\int_0^1 |f(\alpha)|^2 d\alpha = \sum_{p \leqslant n} (\log p)^2 \ll n \log n.$$

Second that, by Theorem 3.1 (cf. the proof of Theorem 2.1),

$$\sup_{\alpha \in \mathfrak{m}} |f(\alpha)| \ll n(\log n)^{4 - B/2}.$$

Thus

Theorem 3.2 *Suppose that A is a positive constant and $B \geqslant 2A + 10$. Then*

$$\int_{\mathfrak{m}} |f(\alpha)|^3 \mathrm{d}\alpha \ll n^2 (\log n)^{-A}.$$

The treatment of the major arcs, although straightforward, requires an appeal to the theory of the distribution of primes in arithmetic progressions.

Lemma 3.1 *Let*

$$v(\beta) = \sum_{m=1}^{n} e(\beta m). \tag{3.6}$$

Then there is a positive constant C such that whenever $1 \leqslant a \leqslant q \leqslant P$, $(a, q) = 1$, $\alpha \in \mathfrak{M}(q, a)$ one has

$$f(\alpha) = \frac{\mu(q)}{\phi(q)} v(\alpha - a/q) + O(n \exp(-C(\log n)^{1/2})).$$

Proof Let

$$f_X(\alpha) = \sum_{p \leqslant X} (\log p) e(\alpha p).$$

Then

$$f_X(a/q) = \sum_{\substack{r=1 \\ (r,q)=1}}^{q} e(ar/q) \vartheta(X, q, r) + O((\log X)(\log q))$$

where

$$\vartheta(X, q, r) = \sum_{\substack{p \leqslant X \\ p \equiv r (\mathrm{mod}\, q)}} \log p.$$

By Theorem 53 and (40) of Estermann (1952) it follows that whenever $\sqrt{n} < X \leqslant n$ one has

$$f_X(a/q) = \frac{X}{\phi(q)} \sum_{\substack{r=1 \\ (r,q)=1}}^{q} e(ar/q) + O(n \exp(-C_1(\log n)^{1/2})). \tag{3.7}$$

Observe that this is trivial when $X \leqslant \sqrt{n}$. Also, by Theorem 271 of Hardy & Wright (1979)

$$\sum_{\substack{r=1 \\ (r,q)=1}}^{q} e(ar/q) = \mu(q).$$

Hence, by (3.1), (3.6), (3.7) and Lemma 2.6 with $X = n$. $F(m) = e(\beta m)$, $\beta = \alpha - a/q$,

$$c_m = \begin{cases} e(am/q)\log m - \mu(q)/\phi(q) & \text{when } m \text{ is prime,} \\ -\mu(q)/\phi(q) & \text{otherwise,} \end{cases}$$

one has

$$f(\alpha) - \frac{\mu(q)}{\phi(q)} v(\alpha - a/q) \ll (1 + n|\alpha - a/q|) n \exp(-C_1(\log n)^{1/2}).$$

With (3.3) and (3.2) this establishes the lemma.

Let $\alpha \in \mathfrak{M}(q,a)$. Then, by the above lemma,

$$f(\alpha)^3 - \frac{\mu(q)}{\phi(q)^3} v(\alpha - a/q)^3 \ll n^3 \exp(-C(\log n)^{1/2}).$$

Now integrating over $\mathfrak{M}$ gives

$$\sum_{q \leqslant P} \sum_{\substack{a=1 \\ (a,q)=1}}^{q} \int_{\mathfrak{M}(q,a)} \left(f(\alpha)^3 - \frac{\mu(q)}{\phi(q)^3} v(\alpha - a/q)^3\right) e(-\alpha n) \mathrm{d}\alpha$$
$$\ll P^3 n^2 \exp(-C(\log n)^{1/2}).$$

Therefore, by (3.3),

$$\int_{\mathfrak{M}} f(\alpha)^3 e(-\alpha n) \mathrm{d}\alpha = \mathfrak{S}(n, P) \int_{-P/n}^{P/n} v(\beta)^3 e(-\beta n) \mathrm{d}\beta + O(P^3 n^2 \exp(-C(\log n)^{1/2})) \tag{3.8}$$

where

$$\mathfrak{S}(n,P) = \sum_{q \leqslant P} \sum_{\substack{a=1 \\ (a,q)=1}}^{q} \frac{\mu(q)}{\phi(q)^3} e(-an/q). \tag{3.9}$$

By (3.6), when β is not an integer,

$$v(\beta) \ll \|\beta\|^{-1}. \tag{3.10}$$

Hence the interval of integration $[-P/n, P/n]$ can be replaced by $[-\frac{1}{2}, \frac{1}{2}]$ with a total error

$$\ll \sum_{q \leqslant P} \phi(q)^{-2} n^2 P^{-2}.$$

Therefore, by (3.2),

$$\int_{\mathfrak{M}} f(\alpha)^3 e(-\alpha n) \mathrm{d}\alpha = \mathfrak{S}(n, P) J(n) + O(n^2 (\log n)^{-2B}) \tag{3.11}$$

where

$$J(n) = \int_{-1/2}^{1/2} v(\beta)^3 e(-\beta n)\mathrm{d}\beta.$$

By (3.6), $J(n)$ is the number of solutions of $m_1 + m_2 + m_3 = n$ with $1 \leqslant m_j \leqslant n$. Thus

$$J(n) = \tfrac{1}{2}(n-1)(n-2). \tag{3.12}$$

Also, by (3.9),

$$\mathfrak{S}(n,P) = \mathfrak{S}(n) + O\left(\sum_{q>P} \phi(q)^{-2}\right)$$

where

$$\mathfrak{S}(n) = \sum_{q=1}^{\infty} \frac{\mu(q)}{\phi(q)^3} \sum_{\substack{a=1\\(a,q)=1}}^{q} e(-an/q). \tag{3.13}$$

Hence, by (3.1), (3.11) and Theorem 327 of Hardy & Wright (1979)

$$\int_{\mathfrak{M}} f(\alpha)^3 e(-\alpha n)\mathrm{d}\alpha = \mathfrak{S}(n)J(n) + O(n^2(\log n)^{-B/2}).$$

By Theorems 67 and 272 of Hardy & Wright (1979) Ramanujan's sum,

$$c_q(n) = \sum_{\substack{a=1\\(a,q)=1}}^{q} e(-an/q)$$

is a multiplicative function of q and satisfies

$$c_q(n) = \frac{\mu(q/(q,n))\phi(q)}{\phi(q/(q,n))}. \tag{3.14}$$

Hence, by (3.13),

$$\mathfrak{S}(n) = \left(\prod_{p|n}(1+(p-1)^{-3})\right)\prod_{p\nmid n}(1-(p-1)^{-2}). \tag{3.15}$$

This establishes

Theorem 3.3 *Suppose that A is a positive constant and $B \geqslant 2A$. Then*

$$\int_{\mathfrak{M}} f(\alpha)^3 e(-\alpha n)\mathrm{d}\alpha = \tfrac{1}{2}n^2\mathfrak{S}(n) + O(n^2(\log n)^{-A})$$

where $\mathfrak{S}(n)$ satisfies (3.15).

Note that $\mathfrak{S}(n) \gg 1$ when n is odd and $\mathfrak{S}(n) = 0$ when n is even. When coupled with Theorem 3.2 and (3.4), Theorem 3.3 yields

Theorem 3.4 *Suppose that A is a positive constant and $R(n)$ satisfies* (3.5). *Then*

$$R(n) = \tfrac{1}{2}n^2\mathfrak{S}(n) + O(n^2(\log n)^{-A})$$

where $\mathfrak{S}(n)$ satisfies (3.15).

Corollary *Every sufficiently large odd number is the sum of three primes.*

3.2 The binary Goldbach problem

In the binary Goldbach problem it is not possible to obtain an asymptotic formula in the same manner as in §3.1. However, a non-trivial estimate can be obtained for

$$\sum_{m=1}^{n} (R_1(m) - m\mathfrak{S}_1(m))^2$$

where

$$R_1(m) = \sum_{\substack{p_1, p_2 \\ p_1 + p_2 = m}} (\log p_1)(\log p_2)$$

and $\mathfrak{S}_1(m)$ is the corresponding singular series. This is because the above expression corresponds to a quaternary problem, rather than to a binary problem. It leads to the less precise conclusion that almost every even number is a sum of two primes.

Let

$$R_1(m) = R_1(m,n) = \sum_{\substack{p_1 \leqslant n \\ p_1 + p_2 = m}} \sum_{p_2 \leqslant n} (\log p_1)(\log p_2). \tag{3.16}$$

Then

$$R_1(m) = R_2(m) + R_3(m) \tag{3.17}$$

where

$$R_2(m) = \int_{\mathfrak{M}} f(\alpha)^2 e(-\alpha m)\,d\alpha \tag{3.18}$$

and

$$R_3(m) = \int_{\mathfrak{m}} f(\alpha)^2 e(-\alpha m)\,\mathrm{d}\alpha. \tag{3.19}$$

Here f, $\mathfrak{M}$, $\mathfrak{m}$ are as in § 3.1.

Now $R_3(m)$ is the Fourier coefficient of the function which is $f(\alpha)^2$ on $\mathfrak{m}$ and 0 elsewhere. Hence, by Bessel's inequality

$$\sum_{m=1}^{n} |R_3(m)|^2 \leqslant \int_{\mathfrak{m}} |f(\alpha)|^4 \mathrm{d}\alpha. \tag{3.20}$$

Theorem 3.5 *Suppose that A is a positive constant and $B \geqslant A+9$. Then*

$$\sum_{m=1}^{n} |R_3(m)|^2 \ll n^3(\log n)^{-A}.$$

This can be deduced, via (3.20), in a similar manner to Theorem 3.2. Let

$$\mathfrak{S}_1(m, P) = \sum_{q \leqslant P} \sum_{\substack{a=1 \\ (a,q)=1}}^{q} \frac{\mu(q)^2}{\phi(q)^2} e(-am/q). \tag{3.21}$$

Then by making only trivial adjustments to the argument that gives (3.8) one obtains

$$R_2(m) = \mathfrak{S}_1(m, P) \int_{-P/n}^{P/n} v(\beta)^2 e(-\beta m)\,\mathrm{d}\beta$$
$$+ O(P^3 n \exp(-C(\log n)^{1/2})).$$

Moreover, by (3.10),

$$\int_{P/n}^{1/2} |v(\beta)|^2 \mathrm{d}\beta \ll nP^{-1}.$$

Hence, by (3.21) and the elementary estimate $\sum_{q \leqslant P} \phi(q)^{-1} \ll \log n$, one has

$$R_2(m) = \mathfrak{S}_1(m,P) J_1(m) + O(n(\log n)^{1-B})$$

where

$$J_1(m) = \int_{-1/2}^{1/2} v(\beta)^2 e(-\beta m)\,\mathrm{d}\beta.$$

By (3.6), $J_1(m)$ is the number of solutions of $m_1 + m_2 = m$ with $1 \leqslant m_j \leqslant n$. Hence, when $m \leqslant n$, one has $J_1(m) = m-1$. Therefore, by (3.21),

$$R_2(m) = m\mathfrak{S}_1(m, P) + O(n(\log n)^{1-B}) \qquad (1 \leqslant m \leqslant n). \tag{3.22}$$

By (3.14), one has

$$\sum_{X<q\leqslant Y}\frac{\mu(q)^2}{\phi(q)^2}\sum_{\substack{a=1\\(a,q)=1}}^{q} e(-am/q)=\sum_{d|m}\frac{\mu(d)^2}{\phi(d)}\sum_{\substack{X/d<q\leqslant Y/d\\(q,m)=1}}\frac{\mu(q)}{\phi(q)^2}$$

$$\ll\sum_{d|m}\frac{\mu(d)^2}{\phi(d)}\min\left(\frac{d}{X},1\right) \tag{3.23}$$

using the elementary fact that

$$\sum_{q>Z}\phi(q)^{-2}\ll Z^{-1}.$$

Hence

$$\mathfrak{S}_1(m)=\sum_{q=1}^{\infty}\frac{\mu(q)^2}{\phi(q)^2}\sum_{\substack{a=1\\(a,q)=1}}^{q} e(-am/q) \tag{3.24}$$

converges,

$$\mathfrak{S}_1(m,P)-\mathfrak{S}_1(m)\ll\log m$$

and

$$\sum_{m=1}^{n}|\mathfrak{S}_1(m,P)-\mathfrak{S}_1(m)|^2\ll(\log n)\sum_{d\leqslant n}\frac{\mu(d)^2 n}{\phi(d)\,d}\min\left(\frac{d}{P},1\right)$$

$$\ll n(\log n)P^{-1}\sum_{d\leqslant n}\frac{\mu(d)^2}{\phi(d)}$$

$$\ll n(\log n)^2P^{-1}.$$

Hence, by (3.2) and (3.22),

$$\sum_{m=1}^{n}|R_2(m)-m\mathfrak{S}_1(m)|^2\ll n^3(\log n)^{2-B}. \tag{3.25}$$

By (3.14),

$$\mathfrak{S}_1(m)=\left(\prod_{p\nmid m}(1-(p-1)^{-2})\right)\prod_{p|m}(1+(p-1)^{-1}). \tag{3.26}$$

Now, by choosing B suitably one obtains

Theorem 3.6 *Suppose that A is a positive constant and $B\geqslant A+2$. Then*

$$\sum_{m=1}^{n}|R_2(m)-m\mathfrak{S}_1(m)|^2\ll n^3(\log n)^{-A}$$

where $\mathfrak{S}_1(m)$ satisfies (3.26).

Combining (3.17) and Theorems 3.5 and 3.6 establishes

Theorem 3.7 *Let A denote a positive constant. Then*

$$\sum_{m=1}^{n} |R_1(m) - m\mathfrak{S}_1(m)|^2 \ll n^3(\log n)^{-A}$$

where R_1 *and* $\mathfrak{S}_1$ *satisfy* (3.16) *and* (3.26) *respectively.*

Note that $\mathfrak{S}_1(m) \gg 1$ when m is even and $\mathfrak{S}_1(m) = 0$ when m is odd.

Corollary *The number $E(n)$ of even numbers m not exceeding n for which m is not the sum of two primes satisfies*

$$E(n) \ll n(\log n)^{-A}.$$

Proof By (3.16) and (3.26), for each m counted by $E(n)$,

$$m^{-2}|R_2(m) - m\mathfrak{S}_1(m)|^2 = \mathfrak{S}_1(m)^2 \gg 1.$$

Hence

$$E(n) \ll \sum_{m=1}^{n} m^{-2}|R_2(m) - m\mathfrak{S}_1(m)|^2.$$

The conclusion now follows from Theorem 3.7 by partial summation.

3.3 Exercises

1 Show that every large natural number can be written in the form $p_1 + p_2 + x^k$.

2 Suppose that $a_1, \ldots, a_4$ are fixed non-zero integers with a_1, a_2, a_3 not all of the same sign. Show that

$$R(n) = \sum_{\substack{p_1 \leqslant n \; p_2 \leqslant n \; p_3 \leqslant n \\ a_1p_1 + a_2p_2 + a_3p_3 + a_4 = 0}} (\log p_1)(\log p_2)(\log p_3)$$

satisfies

$$R(n) = J(n)\mathfrak{S} + O(n^2(\log n)^{-A})$$

where $J(n)$ is the number of solutions of

$$a_1m_1 + a_2m_2 + a_3m_3 + a_4 = 0$$

with $m_j \leqslant n$ and

$$\mathfrak{S} = \sum_{q=1}^{\infty} \phi(q)^{-3} \prod_{j=1}^{4} c_q(a_j).$$

Show that if $(a_1, a_2, a_3)|a_4$, then $J(n) \gg n^2$ for large n.

3 In the notation of the previous exercise show that a sufficient condition for $\mathfrak{S} \gg 1$ to hold is that

$$(a_2, a_3, a_4) = (a_1, a_3, a_4) = (a_1, a_2, a_4) = (a_1, a_2, a_3),$$
$$a_1 + a_2 + a_3 + a_4 \equiv 0 \pmod{2(a_1, a_2, a_3, a_4)}.$$

Show that this condition is also necessary, and that, if it fails, then $\mathfrak{S} = 0$.

4

The major arcs in Waring's problem

4.1 The generating function

The theory of the major arcs in Waring's problem can be refined considerably over that contained in Chapter 2. The intention here is to obtain a relatively good error term for the approximation $V(\alpha, q, a)$ to the generating function $f(\alpha)$ on each major arc whilst making the major arcs as wide and numerous as possible.

Let

$$S(q, a, b) = \sum_{x=1}^{q} e((ax^k + bx)q^{-1}). \tag{4.1}$$

Lemma 4.1 (Hua, 1957*a*) *Suppose that* $(q, a) = 1$. *Then*

$$S(q, a, b) \ll q^{1/2+\varepsilon}(q, b).$$

The proof uses a deep theorem of Weil (see the reference to Schmidt below). There is a more elementary theorem of Davenport & Heilbronn (1936*b*, 1937*a*) in which the exponent $\frac{1}{2}$ is replaced by $\frac{2}{3}$ when $k = 3$ and $\frac{3}{4}$ when $k \geqslant 4$. Also Theorem 7.1 below gives $1 - 1/k$ in place of $\frac{1}{2}$. Indeed the argument of Mordell used to prove Theorem 7.1 in the case when q is prime can be adapted so that when combined with the argument below it gives the Davenport–Heilbronn theorem.

Proof When $(q_1, q_2) = 1$ one has, cf. the proof of Lemma 2.10,

$$S(q_1q_2, a, b) = S(q_1, aq_2^{k-1}, b)S(q_2, aq_1^{k-1}, b).$$

Thus it suffices to show that for each prime power p^l with $p \nmid a$

$$S(p^l, a, b) \ll p^{l/2}(p^l, b). \tag{4.2}$$

When $l = 1$, (4.2) follows at once from Corollary 2F of Chapter II of Schmidt (1976). Thus it can be supposed that

$$l > 1.$$

If $b = 0$, or $b \neq 0$ and the highest power of p, p^θ, dividing b satisfies

$\theta \geqslant l/2$, then (4.2) is trivial. Similarly if the highest power of p, p^{τ}, which divides k satisfies $\tau \geqslant l/2$, then (4.2) is trivial. Thus it can be further assumed that

$$b \neq 0, \ \tau < \tfrac{1}{2}l, \ \theta < \tfrac{1}{2}l.$$

Let

$$\nu = [\tfrac{1}{2}(l+1)].$$

Then $3l - 3\nu \geqslant l$. In the definition of $S(p^l, a, b)$, (4.1), each x, modulo p^l, can be written uniquely in the form $zp^{l-\nu} + y$ with $1 \leqslant y \leqslant p^{l-\nu}$, $1 \leqslant z \leqslant p^{\nu}$. Hence, by the binomial theorem,

$$S(p^l, a, b) = \sum_{y=1}^{p^{l-\nu}} \sum_{z=1}^{p^{\nu}} e\Bigg((ay^k + by)p^{-l} + (kay^{k-1} + b)zp^{-\nu} + \binom{k}{2} ay^{k-2} z^2 p^{l-2\nu}\Bigg). \tag{4.3}$$

Suppose first that l is even or $p|\binom{k}{2}$. Then $\binom{k}{2}p^{l-2\nu}$ is an integer, and hence, by (4.3),

$$|S(p^l, a, b)| \leqslant p^{\nu} N$$

where N is the number of solutions of the congruence

$$kay^{k-1} + b \equiv 0 \pmod{p^{\nu}} \tag{4.4}$$

with $1 \leqslant y \leqslant p^{l-\nu}$. Recall that $\max(\theta, \tau) < l/2 \leqslant \nu$. Thus the congruence is insoluble unless $\theta \geqslant \tau$ and $\theta - \tau$ is a multiple of $k-1$. If (4.4) is insoluble, then (4.2) is immediate. In the contrary case let $\lambda = (\theta - \tau)/(k-1)$. Then N is the number of solutions of

$$(kp^{-\tau})aw^{k-1} + (bp^{-\theta}) \equiv 0 \pmod{p^{\nu-\theta}}$$

with $1 \leqslant w \leqslant p^{l-\nu-\lambda}$. Note that $\lambda \leqslant \theta \leqslant l - \nu$. When $l - \nu - \lambda \leqslant \nu - \theta$, one has $N \ll 1$, so that

$$|S(p^l, a, b)| \ll p^{\nu}.$$

When $l - \nu - \lambda > \nu - \theta$, then $N \ll p^{l+\theta-2\nu-\lambda}$, so that

$$|S(p^l, a, b)| \ll p^{l-\nu} p^{\theta}.$$

In either case

$$|S(p^l, a, b)| \ll p^{\nu}(p^l, b). \tag{4.5}$$

When l is even, $\nu = [(l+1)/2] = l/2$, and when $p|\binom{k}{2}$, $p^{\nu} \leqslant p^{1+l/2} \ll p^{l/2}$. Hence (4.2) follows from (4.5).

It remains to consider the case when l is odd and $p\nmid\binom{k}{2}$. Then

$$v=\tfrac{1}{2}(l+1), \quad v\geqslant 2.$$

Each z in (4.3) can be written uniquely, modulo p^v, in the form $rp+w$ with $1\leqslant r\leqslant p^{v-1}$ and $1\leqslant w\leqslant p$. Moreover

$$\binom{k}{2}ay^{k-2}z^2\equiv\binom{k}{2}ay^{k-2}w^2 \pmod p.$$

Thus the sum over r is zero unless $kay^{k-1}+b\equiv 0 \pmod{p^{v-1}}$. Hence

$$S(p^l,a,b)=p^{v-1}\sum_{y=1}^{p^{l-v}} e((ay^k+by)p^{-l}) \times\sum_{w=1}^{p} e\left(\left(\binom{k}{2}ay^{k-2}w^2+vw\right)p^{-1}\right) \quad (4.6)$$

with y and v satisfying

$$kay^{k-1}+b\equiv 0 \pmod{p^{v-1}} \quad\text{and}\quad v=(kay^{k-1}+b)p^{1-v}. \quad (4.7)$$

First consider the contribution S_1 from those terms with $p\,|\,y^{k-2}$. Then $k>2$ and the innermost sum is zero unless $p\,|\,v$. Thus, by (4.7),

$$S_1\ll p^vN$$

where N is the number of solutions of the congruence

$$kap^{k-1}u^{k-1}+b\equiv 0 \pmod{p^v}$$

with $1\leqslant u\leqslant p^{l-v-1}$. In a similar manner to the treatment of the previous case one obtains $N=0$ unless $\theta=k-1+\tau+(k-1)\lambda$ with $\lambda\geqslant 0$, in which case N is the number of solutions of the congruence

$$(kp^{-\tau})ay^{k-1}+(bp^{-\theta})\equiv 0 \pmod{p^{v-\theta}}$$

with $1\leqslant y\leqslant p^{l-v-1-\lambda}$. Note that $\theta\geqslant k-1>0$. When $l-v-1-\lambda\leqslant v-\theta$ one has $N\ll 1$, and so $S_1\ll p^v\leqslant p^{v-1+\theta}\leqslant p^{l/2}(p^l,b)$. When $l-v-1-\lambda>v-\theta$, then $N\ll p^{l-v-1-\lambda-(v-\theta)}$, so that

$$S_1\ll p^{l-v-1-\lambda+\theta}\leqslant p^{l/2}(p^l,b)$$

once more.

It now remains to estimate the contribution S_2 from those terms in (4.6) with $p\nmid y^{k-2}$. Then the innermost sum is easily seen to be $\ll p^{1/2}$ (cf. the case $k=2$ of Theorem 4.2 below). Thus

$$S_2\ll p^{v-1/2}N$$

where N is the number of solutions of the congruence

$$kay^{k-1} + b \equiv 0 \pmod{p^{\nu-1}}$$

with $1 \leqslant y \leqslant p^{l-\nu}$. Note that $\nu - \frac{1}{2} = \frac{1}{2}l$, $l - \nu = \nu - 1 = \frac{1}{2}(l-1) \geqslant \theta$ and $l \geqslant 3$. If $\theta = \frac{1}{2}(l-1)$, then at once

$$S_2 \ll p^{l/2}(p^l, b).$$

If $\theta < \frac{1}{2}(l-1)$, then as before either $N = 0$, or $\theta - \tau = \lambda(k-1)$ with $\lambda \geqslant 0$, and so $N \ll p^{(l-1)/2 - \lambda - ((l-1)/2 - \theta)} \leqslant p^{\theta}$. Thus in this case also,

$$S_2 \ll p^{l/2}(p^l, b).$$

The next lemma is often the igniting spark for the estimation of exponential sums. It is a truncated form of the Poisson summation formula.

Lemma 4.2 *Suppose that $X < Y$, F'' exists and is continuous on $[X, Y]$ and F' is monotonic on $[X, Y]$. Let H_1, H_2 denote integers such that $H_1 \leqslant F'(\alpha) \leqslant H_2$ for every α in $[X, Y]$. Then*

$$\sum_{X < x \leqslant Y} e(F(x)) = \sum_{h=H_1}^{H_2} \int_X^Y e(F(\alpha) - \alpha h)\alpha\alpha + O(\log(2 + H))$$

where $H = \max(|H_1|, |H_2|)$.

Proof For a differentiable function $\psi(\alpha)$ with ψ' continuous, the Euler–Maclaurin summation formula gives

$$\sum_{X < x \leqslant Y} \psi(x) = \int_X^Y \psi(\alpha)\mathrm{d}\alpha - [\psi(\alpha)(\alpha - [\alpha] - \tfrac{1}{2})]_X^Y$$
$$+ \int_X^Y \psi'(\alpha)(\alpha - [\alpha] - \tfrac{1}{2})\mathrm{d}\alpha. \tag{4.8}$$

Therefore

$$\sum_{X < x \leqslant Y} e(F(x)) = \int_X^Y e(F(\alpha))\mathrm{d}\alpha$$
$$+ \int_X^Y 2\pi i F'(\alpha) e(F(\alpha))(\alpha - [\alpha] - \tfrac{1}{2})\mathrm{d}\alpha + O(1).$$

Now recall the Fourier expansion

$$\alpha - [\alpha] - \tfrac{1}{2} = \sum_{\substack{h=-\infty \\ h \neq 0}}^{\infty} \frac{e(-\alpha h)}{2\pi i h}.$$

This is boundedly convergent for all real α. Hence the second integral above becomes

$$\sum_{\substack{h=-\infty \\ h \neq 0}}^{\infty} \frac{1}{h} \int_X^Y F'(\alpha) e(F(\alpha) - \alpha h) \mathrm{d}\alpha.$$

When $h > H_2$ or $h < H_1$, $F'(\alpha) - h$ is monotonic and non-zero on $[X, Y]$. Hence $F'(\alpha)/(F'(\alpha) - h)$ is also monotonic on $[X, Y]$. Thus integration by parts gives

$$\int_X^Y F'(\alpha) e(F(\alpha) - h\alpha) \mathrm{d}\alpha \ll \left|\frac{F'(Y)}{F'(Y) - h}\right| + \left|\frac{F'(X)}{F'(X) - h}\right|.$$

Therefore

$$\begin{aligned} \sum_{\substack{h=H_2+1 \\ h \neq 0}}^{\infty} \frac{1}{h} \int_X^Y F'(\alpha) e(F(\alpha) - h\alpha) \mathrm{d}\alpha &\ll \sum_{\substack{h=H_2+1 \\ h \neq 0}}^{\infty} \left(\frac{|H_2|}{|h|(h - H_2)} + \frac{|H_1|}{|h|(h - H_1)} \right). \\ &\ll 1 + \sum_{h=1}^{H+1} \frac{1}{h}, \end{aligned}$$

and similarly for the sum over $h \leqslant H_1 - 1$. Integrating the remaining terms by parts gives

$$\sum_{X < x \leqslant Y} e(F(x)) = \int_X^Y e(F(\alpha)) \mathrm{d}\alpha + \sum_{\substack{h=H_1 \\ h \neq 0}}^{H_2} \int_X^Y e(F(\alpha) - \alpha h) \mathrm{d}\alpha + O(\log(2 + H)).$$

If $H_1 \leqslant 0 \leqslant H_2$, then the proof is complete. If $0 < H_1$ or $H_2 < 0$, then $|F'(\alpha)| \geqslant 1$, and so

$$\begin{aligned} \int_X^Y e(F(\alpha)) \mathrm{d}\alpha &= \left[\frac{e(F(\alpha))}{2\pi \mathrm{i} F'(\alpha)} \right]_X^Y + \int_X^Y \frac{F''(\alpha)}{F'(\alpha)^2} \frac{e(F(\alpha))}{2\pi \mathrm{i}} \mathrm{d}\alpha \\ &\ll 1, \end{aligned}$$

which can be absorbed in the error term.

Let

$$f(\alpha) = \sum_{x \leqslant n^{1/k}} e(\alpha x^k), \tag{4.9}$$

$$S(q, a) = \sum_{m=1}^{q} e(am^k/q), \tag{4.10}$$

$$v(\beta) = \sum_{x \leqslant n} \frac{1}{k} x^{1/k-1} e(\beta x), \quad v_1(\beta) = \int_0^{n^{1/k}} e(\beta \gamma^k) d\gamma, \tag{4.11}$$

$$V(\alpha, q, a) = q^{-1} S(q, a) v(\alpha - a/q). \tag{4.12}$$

Theorem 4.1 *Suppose that* $(a, q) = 1$ *and* $\alpha = a/q + \beta$. *Then*

$$f(\alpha) - V(\alpha, q, a) \ll q^{1/2+\varepsilon}(1 + n|\beta|). \tag{4.13}$$

If further $|\beta| \leqslant (2kq)^{-1} n^{1/k-1}$, *then*

$$f(\alpha) - V(\alpha, q, a) \ll q^{1/2+\varepsilon}. \tag{4.14}$$

Also, the same conclusions hold if $v(\beta)$ *is replaced by* $v_1(\beta)$.

The proof uses Lemma 4.1. If any of the weaker results mentioned in the remark after that lemma are used instead, then Theorem 4.1 is obtained with the exponent $\frac{1}{2}$ replaced by the corresponding larger exponent.

Proof Suppose that $X \leqslant n^{1/k}$ and write

$$f_X(\alpha) = \sum_{x \leqslant X} e(\alpha x^k).$$

By (4.1) and (4.9),

$$f_X(\alpha) = \sum_{x \leqslant X} e(\beta x^k) \sum_{\substack{m=1 \\ m \equiv x \pmod q}}^{q} e(am^k/q)$$

$$= q^{-1} \sum_{b=1}^{q} \left(\sum_{x \leqslant X} e(\beta x^k - bx/q) \right) S(q, a, b).$$

Hence

$$f_X(\alpha) - q^{-1} S(q, a) F(q) = q^{-1} \sum_{b=1}^{q-1} F(b) S(q, a, b) \tag{4.15}$$

with

$$F(b) = \sum_{x \leqslant X} e(\beta x^k - bx/q). \tag{4.16}$$

When $\beta = 0$ and $q \nmid b$, $F(b) \ll \|b/q\|^{-1}$. Hence, by Lemma 4.1 and (4.15),

$$f_X(a/q) - q^{-1} S(q, a)[X] \ll q^{-1} \sum_{b=1}^{q-1} \|b/q\|^{-1} q^{1/2+\varepsilon}(q, b)$$

$$\ll q^{1/2+2\varepsilon}.$$

Now (4.13) follows easily by partial integration, cf. the proof of Lemma 2.7. The equivalence of v and v_1 in (4.13) follows from the remark after that lemma.

It remains to prove (4.14). Henceforth suppose that $X = n^{1/k}$, so that in (4.15), $f_X(\alpha) = f(\alpha)$. When $1 \leqslant b \leqslant q$, the expression $\beta k\gamma^{k-1} - b/q$ is a monotonic function of γ on $[0, X]$ and lies between $-(b+\frac{1}{2})/q$ and $-(b-\frac{1}{2})/q$. Thus Lemma 4.2 can be applied with $H_1 = -2$, $H_2 = 0$, whence

$$F(b) = \sum_{h=-2}^{0} \int_0^X e(\beta\gamma^k - b\gamma/q - \gamma h)\mathrm{d}\gamma + O(1). \tag{4.17}$$

When $1 \leqslant b \leqslant q-1$ and $0 \leqslant \gamma \leqslant X$ one has $|\beta k\gamma^{k-1} - b/q - h| \geqslant \|\beta k\gamma^{k-1} - b/q\| \geqslant \frac{1}{2}\|b/q\|$. Hence, by integration by parts,

$$\int_0^X e(\beta\gamma^k - b\gamma/q - \gamma h)\mathrm{d}\gamma \ll \|b/q\|^{-1},$$

and so, by (4.17), $F(b) \ll \|b/q\|^{-1}$. Therefore, by Lemma 4.1, the right-hand side of (4.15) is

$$\begin{aligned} &\ll q^{-1}\sum_{b=1}^{q-1} \|b/q\|^{-1} q^{1/2+\varepsilon}(q, b) \\ &\ll q^{1/2+2\varepsilon}. \end{aligned}$$

Now consider $F(q)$. When $0 \leqslant \gamma \leqslant X$, one has $|\beta k\gamma^{k-1} \pm 1| \geqslant \frac{1}{2}$. Hence integration by parts gives

$$\int_0^X e(\beta\gamma^k \pm \gamma)\mathrm{d}\gamma \ll 1.$$

Therefore, by (4.11) and (4.17),

$$F(q) = v_1(\beta) + O(1). \tag{4.18}$$

This gives (4.14) with $v(\beta)$ replaced by $v_1(\beta)$.

Let

$$G(Y) = \sum_{m \leqslant Y} \frac{1}{k} m^{1/k-1}.$$

The Euler–Maclaurin summation formula, (4.8), gives

$$G(Y) = Y^{1/k} + C_k + O(Y^{1/k-1}).$$

Hence, by Lemma 2.6 and (4.11),

$$\begin{aligned} v(\beta) &= G(X^k)e(\beta X^k) - 2\pi i\beta \int_1^{X^k} G(\gamma)e(\beta\gamma)\mathrm{d}\gamma \\ &= (X + C_k)e(\beta X^k) - 2\pi i\beta \int_1^{X^k} (\gamma^{1/k} + C_k)e(\beta\gamma)\mathrm{d}\gamma \\ &\quad + O(X^{1-k} + |\beta|X). \end{aligned}$$

Integrating by parts and making a change of variables shows that

$$v(\beta) = v_1(\beta) + O(X^{1-k} + |\beta|X).$$

Now (4.14) follows from the corresponding result with $v(\beta)$ replaced by $v_1(\beta)$, and this completes the proof of the theorem.

4.2 The exponential sum $S(q,a)$

Lemma 4.3 *Suppose that $p \nmid a$. Then*

$$S(p, a) = \sum_{\chi \in \mathscr{A}} \bar{\chi}(a)\tau(\chi) \tag{4.19}$$

where $\mathscr{A}$ denotes the set of non-principal characters χ modulo p for which χ^k is principal, and $\tau(\chi)$ denotes the Gauss sum

$$\sum_{x=1}^{p} \chi(x)e(x/p).$$

Also $|\tau(\chi)| = p^{1/2}$ and card $\mathscr{A} = (k, p-1) - 1$.

Proof Let g denote a primitive root modulo p. Then $\mathscr{A}$ is the set of characters χ_h of the form

$$\chi_h(x) = e\left(\frac{h}{(k, p-1)} \operatorname{ind}_g x\right) \quad (p \nmid x)$$

with $1 \leqslant h < (k, p-1)$. Thus

$$1 + \sum_{\chi \in \mathscr{A}} \chi(x)$$

is the number of solutions in y of the congruence $y^k \equiv x \pmod{p}$. Hence

$$S(p, a) = \sum_{x=1}^{p} e(ax/p)\left(1 + \sum_{\chi \in \mathscr{A}} \chi(x)\right),$$

which gives (4.19). The remaining assertions are trivial.

Let τ and γ be as in (2.24) and (2.25). Note that

$$\gamma \leqslant k \text{ unless } k = p = 2 \text{ in which case } \gamma = 3. \tag{4.20}$$

Lemma 4.4 *Suppose that $p \nmid a$ and $l > \gamma$. Then*

$$S(p^l, a) = \begin{cases} p^{l-1} & \text{when } l \leqslant k, \\ p^{k-1} S(p^{l-k}, a) & \text{when } l > k. \end{cases}$$

Proof Recall that a reduced residue modulo p^l is a kth power residue if and only if it is a kth power residue modulo p^γ. Thus

$$S(p^l, a) = \sum_{\substack{y=1 \\ p \nmid y}}^{p^\gamma} \sum_{z=1}^{p^{l-\gamma}} e(a(zp^\gamma + y^k)p^{-l}) + \sum_{y=1}^{p^{l-1}} e(ap^{k-l}y^k).$$

The innermost sum in the double sum is 0, and the sum on the far right is p^{l-1} when $l \leqslant k$ and

$$p^{k-1} S(p^{l-k}, a)$$

when $l > k$.

Lemma 4.5 *Suppose that $(q, r) = (qr, a) = 1$. Then*

$$S(qr, a) = S(q, ar^{k-1}) S(r, aq^{k-1}).$$

Proof See Lemma 2.10.

Theorem 4.2 *Suppose that $(q, a) = 1$. Then*

$$S(q, a) \ll q^{1-1/k}.$$

Proof When $k = 2$,

$$\begin{aligned} |S(q, a)|^2 &= \sum_{x=1}^{q} \sum_{y=1}^{q} e(a(y^2 - x^2)/q) \\ &= \sum_{x=1}^{q} \sum_{z=1}^{q} e(a(z + 2x)z/q) \\ &= q \sum_{\substack{z=1 \\ q \mid 2z}}^{q} e(az^2/q) \\ &\leqslant 2q. \end{aligned}$$

Hence it can be supposed that $k > 2$. Write $l = uk + v$ with $1 \leqslant v \leqslant k$,

$u \geqslant 0$ and suppose that $p \nmid a$. By Lemma 4.4 and (4.20),

$$S(p^l, a) = p^{(k-1)u} S(p^v, a). \tag{4.21}$$

Consider first the case $v > 1$. If $p > k$, then $\gamma = 1$, so that, by Lemma 4.4,

$$S(p^v, a) = p^{v-1}.$$

If $p \leqslant k$, then trivially

$$|S(p^v, a)| \leqslant kp^{v-1}.$$

Hence, by (4.21),

$$|S(p^l, a)| \leqslant \begin{cases} p^{l-l/k} & (p > k), \\ kp^{l-l/k} & (p \leqslant k). \end{cases} \tag{4.22}$$

Now consider the case $v = 1$. By Lemma 4.3,

$$|S(p^v, a)| \leqslant kp^{1/2} \leqslant kp^{-1/6} p^{1-1/k}.$$

Thus, by (4.21),

$$|S(p^l, a)| \leqslant \begin{cases} p^{l-l/k} & (p > k^6), \\ kp^{l-l/k} & (p \leqslant k^6). \end{cases} \tag{4.23}$$

By (4.22), (4.23) and Lemma 4.5,

$$|S(q, a)| \leqslant q^{1-1/k} \prod_{p \leqslant k^6} k,$$

which gives the theorem.

Lemma 4.6 *Suppose that* $(q, a) = 1$. *Then*

$$V(a/q + \beta, q, a) \ll (q^{-1} \min(n, \|\beta\|^{-1}))^{1/k}.$$

Proof At once by (4.12), Theorem 4.2 and Lemma 2.8.

4.3 The singular series

For each integer h, let

$$S_h(q) = \sum_{\substack{a=1 \\ (q,a)=1}}^{q} (S(q, a) q^{-1})^s e(-ah/q). \tag{4.24}$$

Thus, in the notation of (2.16) and (2.18),

$$S(q) = S_n(q), \quad \mathfrak{S}(n) = \sum_{q=1}^{\infty} S_n(q). \tag{4.25}$$

Lemma 4.7 *Suppose that $s \geqslant 1$ and $l = uk + v$ with $1 \leqslant v \leqslant k$. Then*

$$p^{us} S_h(p^l) \ll \begin{cases} p^{-s/2}(p^{1/2}(p^{l-1}, h) + (p^l, h)) & \text{when } l \equiv 1 \pmod{k}, \\ p^{-s}(p^l, h) & \text{when } l \not\equiv 1 \pmod{k}. \end{cases}$$

Moreover, when $\lambda = l - \max(k, \gamma)$ satisfies $\lambda > 0$ and $p^\lambda \nmid h$, then

$$S_h(p^l) = 0.$$

Proof Suppose first that $p > k$, so that $\gamma = 1$. Write $l = uk + v$ with $1 \leqslant v \leqslant k$. Then, by Lemma 4.4,

$$p^{ls} S_h(p^l) = (p^{u(k-1)})^s \sum_{\substack{a=1 \\ p \nmid a}}^{p^l} S(p^v, a)^s e(-ahp^{-l}). \tag{4.26}$$

Each a can be written uniquely in the form $a = xp^v + y$ with $0 \leqslant x < p^{l-v}$, $1 \leqslant y \leqslant p^v$, $p \nmid y$. The sum over x is 0 unless $p^{l-v} | h$, in which case it is p^{l-v}. In the latter case, the sum over y, when $v > 1$, by Lemma 4.4, is

$$p^{s(v-1)} \sum_{\substack{y=1 \\ p \nmid y}}^{p^v} e(-yhp^{-l})$$

and in modulus this does not exceed $p^{s(v-1)}(p^v, hp^{v-l})$. Thus

$$|S_h(p^l)| \leqslant p^{-us-s}(p^l, h) \qquad (l \not\equiv 1 \pmod{k}).$$

On the other hand, when $v = 1$, the sum over y, by Lemma 4.3, is

$$\sum_{\chi_1 \in \mathscr{A}} \cdots \sum_{\chi_s \in \mathscr{A}} \tau(\chi_1) \ldots \tau(\chi_s) \sum_{y=1}^{p} \bar{\chi}_1 \ldots \bar{\chi}_s(y) e(-yhp^{-l}).$$

When $\chi_1 \ldots \chi_s$ is non-principal, the sum over y here is

$$\chi_1 \ldots \chi_s(hp^{1-l}) \tau(\bar{\chi}_1 \ldots \bar{\chi}_s),$$

and when $\chi_1 \ldots \chi_s$ is principal, it is -1 when $p^l \nmid h$ and $p - 1$ when $p^l | h$. Hence, by Lemma 4.3,

$$S_h(p^l) \ll p^{-us-s/2}(p^{1/2}(p^{l-1}, h) + (p^l, h))$$

as required.

Suppose now that $p \leqslant k$. When $l \leqslant \max(\gamma, k)$ the conclusion is trivial. Hence it can be assumed that $l > \max(\gamma, k)$. Write $l = uk + v$ with $\max(\gamma, k) - k < v \leqslant \max(\gamma, k)$. Then, by Lemma 4.4, (4.26) holds.

Moreover, as above, $S_h(p^l)=0$ unless $p^{l-v}|h$, in which case

$$p^{ls}S_h(p^l)=p^{us(k-1)}p^{l-v}\sum_{\substack{y=1\\p\nmid y}}^{p^v} S(p^v,y)^s e(-yhp^{-l})$$
$$\ll p^{us(k-1)}(p^l,h),$$

since $p\leqslant k$. Thus

$$S_h(p^l)\ll p^{-us-vs}(p^l,h)$$

which is more than is required.

Theorem 4.3 *Suppose that* $s\geqslant 4$. *Then*

$$\mathfrak{S}(n)=\sum_{q=1}^{\infty} S_n(q)$$

converges absolutely and $\mathfrak{S}(n)\geqslant 0$. *Also, when* $s\geqslant \max(5,k+2)$ *one has* $\mathfrak{S}(n)\ll 1$, *and when* $\max(4,k)\leqslant s<\max(5,k+2)$, *one has* $\mathfrak{S}(n)\ll n^{\varepsilon}$.

Proof By Lemma 2.11 and (4.25), $S_n(q)$ is a multiplicative function of q. By Lemma 4.7,

$$\sum_{l=1}^{\infty}|S_n(p^l)|\ll np^{-(s-1)/2}\ll np^{-3/2}.$$

Hence

$$\sum_{q\leqslant Q}|S_n(q)|\leqslant \prod_{p\leqslant Q}(1+C_1p^{-3/2})^n\leqslant C_2^n$$

where C_1 and C_2 depend only on k and s. This gives the absolute convergence of $\mathfrak{S}(n)$. That $\mathfrak{S}(n)$ is non-negative follows from Lemma 2.12.

Let p^{θ} denote the highest power of p dividing n and let $l=uk+v$ with $1\leqslant v\leqslant k$. Then, by Lemma 4.7,

$$\left.\begin{aligned} S_n(p^l)&\ll p^{\omega} && \text{when } l\leqslant \theta+\max(k,\gamma),\\ S_n(p^l)&=0 && \text{when } l>\theta+\max(k,\gamma),\end{aligned}\right\}\tag{4.27}$$

where

$$\omega+us-\min(l,\theta)=\begin{cases}-\tfrac12 s & \text{when } l\leqslant\theta \text{ and } v=1,\\ -\tfrac12(s-1) & \text{when } l>\theta \text{ and } v=1,\\ -s & \text{when } v\neq 1.\end{cases}\tag{4.28}$$

Hence

$$\sum_{l=1}^{\infty}|S_n(p^l)|$$

is $O(p^{-3/2})$ when $\theta = 0$, or $\theta \geqslant 1$ and $s \geqslant \max(5, k+2)$, and is $O(\theta)$ when $\theta \geqslant 1$ and $s \geqslant \max(4, k)$. Therefore in the former case $\mathfrak{S}(n) \ll 1$ and in the latter $\mathfrak{S}(n) \ll d(n)^C$ where C depends only on s and k. Hence result.

Lemma 4.8 *Suppose that* $s \geqslant \max(4, k+1)$. *Then*

$$\sum_{q \leqslant Q} q^{1/k} |S_n(q)| \ll (nQ)^\varepsilon.$$

Proof By (4.27) and (4.28),

$$(p^l)^{1/k} S_n(p^l)$$

is $O(\theta)$, $O(p^{-1})$ or 0 according as $l \leqslant \theta$, $\theta < l \leqslant \theta + \max(k, \gamma)$ or $l > \theta + \max(k, \gamma)$. Hence

$$\sum_{q \leqslant Q} q^{1/k} |S_n(q)| \leqslant \prod_{p \leqslant Q} \left(1 + \sum_{l=1}^{\infty} (p^l)^{1/k} |S_n(p^l)|\right)$$

$$\leqslant d(n)^C \prod_{p \leqslant Q} (1 + C/p)$$

where C depends only on s and k.

4.4 The contribution from the major arcs

Let n denote a large natural number

$$N = [n^{1/k}], \tag{4.29}$$

$$P = N/(2k), \tag{4.30}$$

$$\mathfrak{M}(q, a) = \{\alpha : |\alpha - a/q| \leqslant Pq^{-1}n^{-1}\} \tag{4.31}$$

and write $\mathfrak{M}$ for the union of he $\mathfrak{M}(q, a)$ with $1 \leqslant a \leqslant q \leqslant P$ and $(a, q) = 1$. Then the $\mathfrak{M}(q, a)$ are disjoint and contained in

$$\mathcal{U} = (Pn^{-1}, 1 + Pn^{-1}]. \tag{4.32}$$

Let

$$R_{\mathfrak{M}}(m) = \int_{\mathfrak{M}} f(\alpha)^s e(-\alpha m) \mathrm{d}\alpha. \tag{4.33}$$

Lemma 4.9 *Suppose that* $t \geqslant \max(4, k)$ *and that* $\lambda = 0$ *when* $t \geqslant k+1$ *and* $\lambda = 1/k$ *when* $t = k$. *Let*

$$S_t^*(q) = \sum_{\substack{a=1 \\ (a,q)=1}}^{q} |S(q, a) q^{-1}|^t \tag{4.34}$$

Then

$$\sum_{q \leqslant Q} q^{-\lambda} S_t^*(q) \ll Q^{\varepsilon}.$$

Proof In the same manner as for $S_n(q)$ and with the same notation as in Lemma 4.7 it can be shown that $S_t^*(q)$ is multiplicative and

$$p^{ut-l} S_t^*(p^l)$$

is $O(p^{-t/2})$ when $l \equiv 1 \pmod k$ and is $O(p^{-t})$ when $l \not\equiv 1 \pmod k$. Thus

$$\sum_{q \leqslant Q} q^{-\lambda} S_t^*(q) \leqslant \prod_{p \leqslant Q} \left(1 + \sum_{l=1}^{\infty} p^{-\lambda l} S_t^*(p^l)\right)$$

and

$$\sum_{l=1}^{\infty} p^{-\lambda l} S_t^*(p^l) \ll \sum_{u=0}^{\infty} p^{-u\lambda k + uk - ut}\left(p^{1-\lambda-t/2} + \sum_{v=2}^{k} p^{-v\lambda+v-t}\right)$$
$$\ll p^{-1}$$

provided that $\lambda \geqslant \max((1+k-t)/k, 2-\frac{1}{2}t)$.

Theorem 4.4 *Suppose that $s \geqslant \max(5, k+1)$. Then there is a positive number δ such that whenever $1 \leqslant m \leqslant n$, one has*

$$R_{\mathfrak{M}}(m) = \Gamma\left(1+\frac{1}{k}\right)^s \Gamma\left(\frac{s}{k}\right)^{-1} m^{s/k-1}\mathfrak{S}(m) + O(n^{s/k-1-\delta}).$$

Proof Let $\alpha \in \mathfrak{M}(q, a)$. Then, by Theorem 4.1, (4.29), (4.30) and (4.31),

$$f(\alpha) - V(\alpha, q, a) \ll q^{1/2+\varepsilon}.$$

Therefore

$$f(\alpha)^s - V(\alpha, q, a)^s \ll (q^{1/2+\varepsilon})^s + q^{1/2+\varepsilon}|V(\alpha, q, a)|^{s-1}.$$

Hence, by (4.12), (4.31) and (4.34),

$$\sum_{\substack{a=1 \\ (a,q)=1}}^{q} \int_{\mathfrak{M}(q,a)} (f(\alpha)^s - V(\alpha, q, a)^s) e(-\alpha m) \mathrm{d}\alpha$$
$$\ll Pn^{-1}(q^{1/2+\varepsilon})^s + q^{1/2+\varepsilon} S_{s-1}^*(q) \int_{-1/2}^{1/2} |v(\beta)|^{s-1} \mathrm{d}\beta.$$

Therefore, by (4.33) and Lemma 2.8,

$$R_{\mathfrak{M}}(m) = \sum_{q \leqslant P} \sum_{\substack{a=1 \\ (a,q)=1}}^{q} \int_{\mathfrak{M}(q,a)} V(\alpha, q, a)^s e(-\alpha m)\mathrm{d}\alpha + E \tag{4.35}$$

where, taking into account the possibility $s = k + 1$,

$$E \ll P^2 n^{-1}(P^{1/2+\varepsilon})^s + P^{3/4+\varepsilon} \sum_{q \leqslant P} q^{-1/4} S^*_{s-1}(q) n^{(s-1)/k - 1 + \varepsilon}.$$

By Lemma 4.9, (4.29) and (4.30),

$$E \ll n^{s/k - 1 - \delta}, \tag{4.36}$$

for a suitable positive number δ.

Let $\mathfrak{N}(q, a) = \{\alpha : P/(qn) < |\alpha - a/q| \leqslant \frac{1}{2}\}$. Then, by (4.12), (4.24) and Lemma 2.8,

$$\sum_{\substack{a=1 \\ (a,q)=1}}^{q} \int_{\mathfrak{N}(q,a)} V(\alpha, q, a)^s e(-\alpha m)\mathrm{d}\alpha \ll |S_m(q)| \int_{P/(nq)}^{\infty} \beta^{-s/k}\mathrm{d}\beta \ll (nq/P)^{s/k - 1} |S_m(q)|.$$

Hence, by Lemma 4.8,

$$\sum_{q \leqslant P} \sum_{\substack{a=1 \\ (a,q)=1}}^{q} \int_{\mathfrak{N}(q,a)} V(\alpha, q, a)^s e(-\alpha m)\mathrm{d}\alpha \ll n^{s/k - 1 - \delta}.$$

Therefore, by (4.35), (4.36) and (4.24),

$$R_{\mathfrak{M}}(m) = \mathfrak{S}(m, P) I(m) + O(n^{s/k - 1 - \delta}) \tag{4.37}$$

where

$$\mathfrak{S}(m, P) = \sum_{q \leqslant P} S_m(q), \quad I(m) = \int_{-1/2}^{1/2} v(\beta)^s e(-\beta m)\mathrm{d}\beta.$$

By Lemma 4.8,

$$\sum_{Q < q \leqslant 2Q} |S_m(q)| \ll n^{\varepsilon} Q^{\varepsilon - 1/k}.$$

Hence, by (4.29) and (4.30),

$$\sum_{q > P} |S_m(q)| \ll n^{-\delta}$$

so that, by (4.25),

$$\mathfrak{S}(m, P) = \mathfrak{S}(m) + O(n^{-\delta}). \tag{4.38}$$

Finally, by (4.11), (2.20) and Theorem 2.3, when $1 \leqslant m \leqslant n$,

$$I(m) = J(m)$$
$$= \Gamma\left(1 + \frac{1}{k}\right)^s \Gamma\left(\frac{s}{k}\right)^{-1} m^{s/k - 1}(1 + O(m^{-1/k})).$$

The theorem now follows from Theorem 4.3, (4.37) and (4.38).

4.5 The congruence condition

Let $M_n(q)$ and $M_n^*(q)$ be as in § 2.6.

Theorem 4.5 *Suppose that* $s \geqslant \max(4, k+1)$ *and* $M_n^*(p^\gamma) > 0$ *for every prime p. Then* $\mathfrak{S}(n) \gg 1$.

Proof By Lemmas 2.12 and 2.13,

$$\sum_{h=0}^{\infty} S_n(p^h) \geqslant p^{\gamma(1-s)},$$

the absolute convergence of the infinite series being ensured by that of $\mathfrak{S}(n)$, cf. Theorem 4.3.

It now suffices to show that when $p > k$ one has

$$\sum_{l=1}^{\infty} S_n(p^l) \geqslant -Cp^{-3/2} \tag{4.39}$$

Note that $\gamma = 1$. The argument of Lemma 4.7 shows that if $l = uk + v$ with $2 \leqslant v \leqslant k$, then

$$p^{(u+1)s-l} S_n(p^l) = 1 - \frac{1}{p}, \ -\frac{1}{p}, \ 0$$
$$\text{according as } p^l | n, \ p^{l-1} \| n, \ p^{l-1} \nmid n \tag{4.40}$$

and that if $l \equiv 1 \pmod k$, then $S_n(p^l) = 0$ unless $p^{l-1} | n$ in which case

$$p^{-[l/k](k-s)} S_n(p^l)$$
$$= p^{-s} \sum_{\chi_1 \in \mathscr{A}} \dots \sum_{\chi_s \in \mathscr{A}} \tau(\chi_1) \dots \tau(\chi_s) \sum_{a=1}^{p-1} \bar{\chi}_1 \dots \bar{\chi}_s(a) e(-anp^{-l}). \tag{4.41}$$

Choose θ so that $p^\theta \| n$. Then, by (4.40),

$$\sum_{l \not\equiv 1 \pmod k} S_n(p^l) \geqslant -p^\lambda$$

where $\lambda = [\theta/k](k - s) + 1 - s$. It is readily seen that $\lambda \leqslant -2$.

By Lemma 4.3, the terms in (4.41) with $\chi_1 \ldots \chi_s \neq \chi_0$ contribute $\ll p^{(s+1)/2}$ and if $p \nmid np^{1-l}$, then those with $\chi_1 \ldots \chi_s = \chi_0$ contribute $\ll p^{s/2}$. Hence

$$\sum_{l \equiv 1 \pmod k} S_n(p^l) = \sum_{\substack{l \equiv 1 \pmod k \\ p^l \mid n}} S_n(p^l) + O(p^{-3/2}).$$

If $s \geqslant 5$, then, by Lemma 4.3 and (4.41),

$$S_n(p^l) \ll p^{[l/k](k-s)-3/2}$$

so that

$$\sum_{\substack{l \equiv 1 \pmod k \\ p^l \mid n}} S_n(p^l) \ll p^{-3/2}.$$

Hence it remains to consider $S_n(p^l)$ when $p^l|n$ and $s = 4$.

By (4.41),

$$p^{-[l/k](k-s)} S_n(p^l) = S_{np^{1-l}}(p) = S_p(p).$$

It therefore suffices to show that $S_p(p) \geqslant 0$.

By (4.24),

$$S_p(p) = \sum_{a=1}^{p-1} (S(p, a)p^{-1})^4.$$

The proof is completed by showing that when $k = 2$ or 3 and $p > k$, $S(p, a)$ is real or purely imaginary. Observe that when $s = 4$, one has $k = 2$ or 3.

When $k = 2$, Lemma 4.3 gives

$$S(p, a) = \chi(a)\tau(\chi)$$

where χ is the Legendre symbol. Thus

$$\bar{S}(p, a) = S(p, -a) = \chi(-a)\tau(\chi) = \chi(-1)S(p, a),$$

so that $S(p, a)$ is real or purely imaginary according as $\chi(-1) = 1$ or $\chi(-1) = -1$.

When $k = 3$, one has $(-x)^k = -x^k$. Thus

$$\bar{S}(p, a) = S(p, -a) = S(p, a),$$

so that $S(p, a)$ is real.

Theorem 4.6 *Suppose that* $s \geqslant 5$ *when* $k = 2$, $s \geqslant 4k$ *when* k *is a power of* 2 *with* $k > 2$, *and* $s \geqslant \frac{3}{2}k$ *otherwise. Then* $\mathfrak{S}(n) \gg 1$.

Proof At once from Lemma 2.15 and Theorem 4.5.

4.6 Exercises

1 Show that (4.13) holds with the left hand side replaced by $q^{1/2+\varepsilon}(1+n|\beta|)^{1/2}$. Deduce the special case $\phi(x)=\alpha x^3$ of Lemma 2.4 (Weyl's inequality).

2 Consider the statements

(i) $s \geqslant 4$ and $\mathfrak{S}(n) \gg 1$,

(ii) $M_n(q) > 0$ for every n and for every large q,

(iii) $M_n^*(q) > 0$ for every n and for every large q.

Show that if (i) holds for every n, then (ii) holds, and that if $k \neq 2$ or 4, then (ii) implies (iii).

3 Suppose that $s_0(k)$ is given by the following table.

k	3	4	5	6	7	8	9	10	11	12	13	14	15	16
$s_0(k)$	4	16	5	9	4	32	13	12	11	16	6	4	15	64

Show that when $s \geqslant s_0(k)$ one has $\mathfrak{S}(n) \gg 1$ for every n.

5

Vinogradov's methods

5.1 Vinogradov's mean value theorem

When k is small, i.e. less than 11 or 12 or so, Lemmas 2.4 and 2.5, the essential ingredients for the estimation of the minor arcs in Chapter 2, are the best that are known. However, for larger k, significant improvements can be obtained via Vinogradov's mean value theorem. This theorem is also of importance in the theory of the Riemann zeta function.

In order to describe the theorem it is necessary to introduce some notation. Let $\mathscr{U}_k$ denote the k-dimensional unit hypercube $(0, 1]^k$ and write

$$f(\boldsymbol{\alpha}) = \sum_{Y < x \leqslant Y + X} (\alpha_1 x + \alpha_2 x^2 + \ldots + \alpha_k x^k). \tag{5.1}$$

For each k-tuple $\boldsymbol{h} = (h_1, \ldots, h_k)$ of integers h_j, let $J_s^{(k)}(X, Y, \boldsymbol{h})$ denote the number of solutions of the k simultaneous equations

$$\sum_{r=1}^{s} (x_r^j - y_r^j) = h_j \quad (1 \leqslant j \leqslant k) \text{ with } Y < x_r, y_r \leqslant Y + X. \tag{5.2}$$

Then

$$J_s^{(k)}(X, Y, \boldsymbol{h}) = \int_{\mathscr{U}_k} |f(\boldsymbol{\alpha})|^{2s} e(\boldsymbol{\alpha} \cdot \boldsymbol{h}) \mathrm{d}\boldsymbol{\alpha} \tag{5.3}$$

where $\boldsymbol{\alpha} \cdot \boldsymbol{h}$ denotes the scalar product $\alpha_1 h_1 + \ldots + \alpha_k h_k$. Trivially

$$J_s^{(k)}(X, Y, \boldsymbol{h}) \leqslant J_s^{(k)}(X, Y, \boldsymbol{0}). \tag{5.4}$$

By writing $x_r = M + u_r$, $y_r = M + v_r$ and applying the binomial theorem, it is easily seen that

$$J_s^{(k)}(N, M, \boldsymbol{0}) = J_s^{(k)}(N, 0, \boldsymbol{0}). \tag{5.5}$$

For brevity write

$$J_s(X) = J_s^{(k)}(X) = J_s^{(k)}(X, 0, \boldsymbol{0}). \tag{5.6}$$

By (5.2), $J_s(X)$ is the number of solutions of

$$\sum_{r=1}^{s} (x_r^j - y_r^j) = 0 \quad (1 \leqslant j \leqslant k) \quad \text{with } 0 < x_r, y_r \leqslant X. \tag{5.7}$$

A non-trivial estimate for $J_s(X)$ is known as 'Vinogradov's mean value theorem'.

All known methods for estimating $J_s(X)$ when k is large depend on a reduction which relates $J_s(X)$ to $J_{s-k}(X/p)$ where p is a suitable prime number. The method adopted here is a refinement by Bombieri of an argument of Karatsuba.

Lemma 5.1 (Karatsuba) *Suppose that p is a prime number with $p > k$. Let $A(p, \mathbf{h})$ denote the number of solutions of the k simultaneous congruences*

$$\sum_{r=1}^{k} n_r^j \equiv h_j (\mathrm{mod}\, p^j)\,(1 \leqslant j \leqslant k)$$

with $n_r \leqslant p^k$ and the n_r distinct modulo p. Then

$$A(p, \mathbf{h}) \leqslant k!\,p^{k(k-1)/2}.$$

Proof Let $B(\mathbf{g})$ denote the number of solutions of

$$\sum_{r=1}^{k} n_r^j \equiv g_j (\mathrm{mod}\, p^k) \qquad (1 \leqslant j \leqslant k) \tag{5.8}$$

with $n_r \leqslant p^k$ and the n_r distinct modulo p. Then $A(p, \mathbf{h})$ is the sum of all the $B(\mathbf{g})$ with $g_j \equiv h_j (\mathrm{mod}\, p^j)$ and $1 \leqslant g_j \leqslant p^k (1 \leqslant j \leqslant k)$. The total number of possible choices for $\mathbf{g}$ is $p^{k(k-1)/2}$. Thus it now suffices to show that

$$B(\mathbf{g}) \leqslant k!$$

and this will follow on showing that every solution of (5.8) is a permutation of any given solution.

For a given $\mathbf{g}$, suppose that $n_1, \ldots, n_k$ is a solution of (5.8) with $n_r \leqslant p^k$ and the n_r distinct modulo p. Suppose that $m_1, \ldots, m_k$ is another such solution and let

$$P(x) = \prod_{r=1}^{k} (x - n_r). \tag{5.9}$$

Then, by Newton's formulae connecting the sums of the powers of the roots of a polynomial with its coefficients, and the fact that $p > k$, it follows that

$$P(x) \equiv \prod_{r=1}^{k} (x - m_r)\,(\mathrm{mod}\, p^k).$$

Thus

$$P(m_r) \equiv 0 \pmod{p^k} \quad (1 \leqslant r \leqslant k). \tag{5.10}$$

Hence, for each r there is an s such that $n_s \equiv m_r \pmod p$. Also, since the n_s are distinct modulo p it follows that n_s is unique. Hence, by (5.9) and (5.10), $n_s \equiv m_r \pmod{p^k}$, whence $n_s = m_r$. Thus the m_r are a permutation of the n_r, as required.

Let p denote a prime number, let $R_1(\boldsymbol{h})$ denote the number of solutions of the simultaneous equations

$$\sum_{r=1}^{s} x_r^j = h_j \quad (1 \leqslant j \leqslant k)$$

with $0 < x_r \leqslant X$ and at least k of the x_r distinct modulo p, and let $R_2(\boldsymbol{h})$ denote the corresponding number with at most $k-1$ of the x_r distinct modulo p. Then

$$J_s(X) = \sum_{\boldsymbol{h}} (R_1(\boldsymbol{h}) + R_2(\boldsymbol{h}))^2$$

$$\leqslant 2 \sum_{\boldsymbol{h}} (R_1(\boldsymbol{h})^2 + R_2(\boldsymbol{h})^2).$$

Hence

$$J_s(X) \leqslant 2I_1(p) + 2I_2(p) \tag{5.11}$$

where $I_1(p)$ is the number of solutions of (5.7) with at least k of the x_r distinct modulo p and at least k of the y_r distinct modulo p, and where $I_2(p)$ is the number of solutions of (5.7) with neither the x_r taking on more than $k-1$ distinct values modulo p nor likewise the y_r.

The solutions of (5.7) counted by $I_1(p)$ can be typified by those in which $x_1, \ldots, x_k$ are distinct modulo p and $y_1, \ldots, y_k$ are distinct modulo p. Hence, on writing I_3 for the number of such solutions, it is immediate that

$$I_1(p) \leqslant \binom{s}{k}^2 I_3. \tag{5.12}$$

Let

$$f(\boldsymbol{\alpha}, y) = \sum_{\substack{0 < x \leqslant X \\ x \equiv y \pmod p}} e(\alpha_1 x + \alpha_2 x^2 + \ldots + \alpha_k x^k),$$

and let $\mathscr{A}$ denote the set of k-tuples $\boldsymbol{a} = (a_1, \ldots, a_k)$ with $0 < a_r \leqslant p$ and the a_r distinct. Then

$$I_3 = \int_{\mathscr{U}_k} \left| \sum_{x \leqslant p} f(\boldsymbol{\alpha}, x) \right|^{2s-2k} \left| \sum_{a \in \mathscr{A}} f(\boldsymbol{\alpha}, a_1) \ldots f(\boldsymbol{\alpha}, a_k) \right|^2 \mathrm{d}\boldsymbol{\alpha}.$$

By Hölder's inequality

$$\left| \sum_{x \leqslant p} f(\boldsymbol{\alpha}, x) \right|^{2s-2k} \leqslant p^{2s-2k-1} \sum_{x \leqslant p} \left| f(\boldsymbol{\alpha}, x) \right|^{2s-2k}$$

Hence

$$I_3 \leqslant p^{2s-2k} \max_{x \leqslant p} I_4(x) \tag{5.13}$$

where $I_4(x)$ is the number of solutions of the simultaneous equations

$$\sum_{r=1}^{k} (m_r^j - n_r^j) = \sum_{r=1}^{s-k} ((py_r + x)^j - (pz_r + x)^j) \quad (1 \leqslant j \leqslant k)$$

with $0 < m_r, n_r \leqslant X$, with $-x/p < y_r, z_r \leqslant (X - x)/p$, with $m_1, \ldots, m_k$ distinct modulo p, and with $n_1, \ldots, n_k$ distinct modulo p. A simple application of the binomial theorem shows that $I_4(x)$ is the number of solutions of the simultaneous equations

$$\sum_{r=1}^{k} ((m_r - x)^j - (n_r - x)^j) = \sum_{r=1}^{s-k} p^j (y_r^j - z_r^j) \quad (1 \leqslant j \leqslant k)$$

with the variables satisfying the same conditions as before.

Now suppose that

$$p^k \geqslant X, \quad p > k. \tag{5.14}$$

Then, by Lemma 5.1, (5.4), (5.5) and (5.6),

$$\begin{aligned} I_4(x) &\leqslant X^k k!\, p^{k(k-1)/2} \max_{\mathbf{h}} J_{s-k}^{(k)}(X/p, -x/p, \mathbf{h}) \\ &\leqslant X^k k!\, p^{k(k-1)/2} J_{s-k}(1 + Xp^{-1}). \end{aligned}$$

This together with (5.11), (5.12) and (5.13) gives

$$\begin{aligned} J_s(X) &\leqslant 2I_2(p) \\ &\quad + 2\binom{s}{k} k!\, p^{2s + k(k-5)/2} X^k J_{s-k}(1 + Xp^{-1}). \end{aligned} \tag{5.15}$$

Lemma 5.2 (Bombieri) *Let* $\lambda > 0$, *and let* $\mathscr{B}_1, \mathscr{B}_2, \ldots, \mathscr{B}_s$ *denote* s *subsets of a finite set* $\mathscr{B}$ *with*

$$\operatorname{card} \mathscr{B}_r \geqslant \lambda \operatorname{card} \mathscr{B} \quad (1 \leqslant r \leqslant s).$$

Then for every $t < \lambda s$ there exist $r_1, r_2, \ldots, r_t$ with $r_1 < r_2 < \ldots < r_t$ and

$$\operatorname{card}(\mathscr{B}_{r_1} \cap \mathscr{B}_{r_2} \cap \ldots \cap \mathscr{B}_{r_t}) \geqslant \left(\lambda - \frac{t}{s}\right)\binom{s}{t}^{-1} \operatorname{card}\mathscr{B}.$$

Proof Let $\mathscr{C}$ denote the set of elements b of $\mathscr{B}$ for which b belongs to at least t of the $\mathscr{B}_r$. Then

$$s\lambda \operatorname{card}\mathscr{B} \leqslant \sum_{r=1}^{s} \operatorname{card}\mathscr{B}_r \leqslant t \operatorname{card}\mathscr{B} + s \operatorname{card}\mathscr{C},$$

whence

$$\operatorname{card}\mathscr{C} \geqslant \left(\lambda - \frac{t}{s}\right)\operatorname{card}\mathscr{B}.$$

Moreover

$$\operatorname{card}\mathscr{C} \leqslant \sum_{\substack{r_1, \ldots, r_t \\ r_1 < r_2 < \ldots < r_t}} \operatorname{card}(\mathscr{B}_{r_1} \cap \mathscr{B}_{r_2} \cap \ldots \cap \mathscr{B}_{r_t})$$

and the number of terms in the multiple sum is $\binom{s}{t}$. Choosing $r_1, \ldots, r_t$ to correspond to a maximal term gives the desired conclusion.

Theorem 5.1 (Vinogradov's mean value theorem) *For each pair of natural numbers k, l there exists a positive number $C(l, k)$ such that for every $X > 0$*

$$J_{lk}^{(k)}(X) \leqslant C(l, k) X^{2lk - k(k+1)/2 + \eta}$$

where $\eta = \frac{1}{2}k^2(1 - 1/k)^l$.

It should be observed that for applications in multiplicative number theory it is necessary to know something of the behaviour of $C(l, k)$ as l and k grow. Here, however, it is of lesser importance. Note that the theorem is trivial when $k = 1$.

Proof This is by induction on l. By a similar argument to that used to estimate $B(\mathbf{g})$ in the proof of Lemma 5.1, it can be shown that when $s = k$ all the solutions of (5.7) are obtained with the y_r as permutations of the x_r. Thus

$$J_k(X) \leqslant k! X^k$$

which gives the case $l = 1$ at once.

Now suppose that $l > 1$ and the theorem holds with l replaced by $l-1$. When $X \leqslant k^k$ the desired conclusion is trivial. Thus it may be assumed that $X > k^k$. Let p be any prime number with $X^{1/k} \leqslant p \leqslant 2^{5k}X^{1/k}$, so that (5.14) holds. Then, by (5.15) and the inductive hypothesis,

$$J_{kl}(X) \leqslant 2I_2(p) \\ + C_1(k, l)p^{2kl + k(k-5)/2}(X/p)^{2kl - k(k+5)/2 + \eta'}X^k$$

where $\eta' = \frac{1}{2}k^2(1-1/k)^{l-1}$. The exponent of p here is $k^2 - \eta'$, so that

$$J_{kl}(X) \leqslant 2I_2(p) + C_2(k, l)X^{2kl - k(k+1)/2 + \eta}.$$

The proof now divides into cases. The first case occurs when there is at least one prime p in the range $X^{1/k} \leqslant p \leqslant 2^{5k}X^{1/k}$ for which

$$I_2(p) \leqslant \tfrac{1}{4}J_{kl}(X).$$

Then the theorem follows at once.

The second case is the contrary one in which there are no such primes. Then, by Bertrand's postulate, there are at least $5k$ distinct primes $p_1, \ldots, p_{5k}$ such that

$$I_2(p_r) > \tfrac{1}{4}J_{kl}(X) \tag{5.16}$$

and

$$X^{1/k} \leqslant p_r \leqslant 2^{5k}X^{1/k}. \tag{5.17}$$

Let $\mathscr{B}$ denote the set of solutions to (5.7) with $s = kl$. Then

$$\operatorname{card}\mathscr{B} = J_{kl}(X).$$

Let $\mathscr{B}(p)$ denote the subset of $\mathscr{B}$ consisting of the solutions in which neither the x_r take on more than $k-1$ distinct values modulo p nor do the y_r. Then

$$\operatorname{card}\mathscr{B}(p_r) = I_2(p_r).$$

Thus, by (5.16) and Lemma 5.2 with $\lambda = \frac{1}{4}$, $s = 5k$, $t = k$, there are k distinct primes $q_1, \ldots, q_k$ such that

$$J_{kl}(X) \leqslant 20\tbinom{5k}{k}\operatorname{card}(\mathscr{B}(q_1)\cap\mathscr{B}(q_2)\cap\ldots\cap\mathscr{B}(q_k)). \tag{5.18}$$

Now consider a typical element of $\mathscr{B}(q_1)\cap\mathscr{B}(q_2)\cap\ldots\cap\mathscr{B}(q_k)$. Since $x_r \leqslant X$ and $q_1 \ldots q_k \geqslant X$ it follows that each x_r is uniquely determined by its residue classes modulo q_1, modulo $q_2, \ldots,$ modulo

q_k. Moreover $x_1, \ldots, x_{kl}$ lie in at most $k-1$ residue classes modulo q_t. Thus the number of choices for the kl-tuple $(x_1, \ldots, x_{kl})$ modulo q_t is at most $q_t^{k-1}(k-1)^{kl}$. Thus the number of choices modulo $q_1 \ldots q_k$ is at most $(q_1 \ldots q_k)^{k-1}(k-1)^{k^2 l}$. Similarly for $(y_1, \ldots, y_{kl})$. Therefore

$$\operatorname{card}(\mathscr{B}(q_1) \cap \mathscr{B}(q_2) \cap \ldots \cap \mathscr{B}(q_k)) \leqslant (q_1 \ldots q_k)^{2k-2}(k-1)^{2k^2 l}.$$

Hence, by (5.18),

$$J_{kl}(X) \leqslant C_3(k, l) X^{2k-2}.$$

On the other hand the trivial solutions of (5.7) with $s = kl$ contribute at least $[X]^{kl} \geqslant [X]^{2k}$ to $J_{kl}(X)$ when $l > 1$. Thus $X \leqslant C_4(k, l)$ which gives the conclusion in the second case.

5.2 The transition from the mean

Let

$$\mathbf{v}(x) = \mathbf{v}^{(k)}(x) = (x, x^2, \ldots, x^k) \tag{5.19}$$

and for $\boldsymbol{\alpha} \in \mathbb{R}^k$ write

$$f(\boldsymbol{\alpha}) = \sum_{x=1}^{N} e(\mathbf{v}(x) \cdot \boldsymbol{\alpha}) \tag{5.20}$$

where as usual for two elements $\boldsymbol{\alpha}$, $\boldsymbol{\beta}$ of $\mathbb{R}^k$, $\boldsymbol{\alpha} \cdot \boldsymbol{\beta}$ denotes the scalar product $\alpha_1\beta_1 + \ldots + \alpha_k\beta_k$. Suppose that $\mathscr{M}$ is a non-empty set of integers with

$$\mathscr{M} \subset [1, N], \quad M = \operatorname{card}(\mathscr{M}). \tag{5.21}$$

Then, for $m \in \mathscr{M}$,

$$\begin{aligned} f(\boldsymbol{\alpha}) &= \sum_{x=1+m}^{N+m} e(\mathbf{v}(x-m) \cdot \boldsymbol{\alpha}) \\ &= \int_0^1 \left(\sum_{x=1}^{2N} e(\mathbf{v}(x-m) \cdot \boldsymbol{\alpha} + x\beta) \right) \sum_{y=1+m}^{N+m} e(-y\beta) \mathrm{d}\beta. \end{aligned}$$

Therefore, on summing over the elements of $\mathscr{M}$,

$$f(\boldsymbol{\alpha}) \ll M^{-1} \int_0^1 \left(\sum_{m \in \mathscr{M}} |g(m, \beta)| \right) \min(N, \|\beta\|^{-1}) \mathrm{d}\beta$$

where

$$g(m, \beta) = \sum_{x=1}^{2N} e(\mathbf{v}(x-m) \cdot \boldsymbol{\alpha} + x\beta). \tag{5.22}$$

Thus

$$f(\boldsymbol{\alpha}) \ll M^{-1}(\log 2N) \sup_{0 \leqslant \beta \leqslant 1} \sum_{m \in \mathcal{M}} |g(m, \beta)|, \tag{5.23}$$

and an estimate for f can be deduced from a suitable mean value theorem.

The following lemma embodies a relationship between discrete mean values and corresponding continuous mean values. Note that

$$\sum_{\boldsymbol{n} \in \mathcal{N}} |a(\boldsymbol{n})|^2 = \int_{\mathcal{U}_k} |S(\boldsymbol{\beta})|^2 \mathrm{d}\boldsymbol{\beta}.$$

A number of alternative methods have been devised for obtaining such relationships, but all are based on similar ideas. The method given here is suggested by the large sieve inequality, and is a generalization of that inequality to k dimensions which has been useful in algebraic number theory. See Huxley (1968) and Wilson (1969).

Lemma 5.3 *Suppose that* $\delta_j > 0$ $(j = 1, \ldots, l)$ *and that* Γ *is a non-empty set of points* γ *in* $\mathbb{R}^l$ *such that the open sets*

$$\mathcal{R}(\gamma) = \{\boldsymbol{\beta} : \|\beta_j - \gamma_j\| < \delta_j,\ 0 \leqslant \beta_j < 1\}$$

are pairwise disjoint. Let $N_1, \ldots, N_l$ *denote* l *natural numbers and* $\mathcal{N}$ *denote the set of integer* l*-tuples* $\boldsymbol{n} = (n_1, \ldots, n_l)$ *with* $1 \leqslant n_j \leqslant N_j$. *Then the sum*

$$S(\boldsymbol{\beta}) = \sum_{\boldsymbol{n} \in \mathcal{N}} a(\boldsymbol{n}) e(\boldsymbol{n} \cdot \boldsymbol{\beta}),$$

where the $a(\boldsymbol{n})$ *are complex numbers, satisfies*

$$\sum_{\gamma \in \Gamma} |S(\gamma)|^2 \ll \sum_{\boldsymbol{n} \in \mathcal{N}} |a(\boldsymbol{n})|^2 \prod_{j=1}^{l} (N_j + \delta_j^{-1}).$$

Proof Without loss of generality it can be supposed that $0 < \delta_j < 1$. It suffices to bound the dual form

$$\sum_{\boldsymbol{n} \in \mathcal{N}} |T(\boldsymbol{n})|^2$$

where

$$T(\boldsymbol{n}) = \sum_{\gamma \in \Gamma} b(\gamma) e(\boldsymbol{n} \cdot \gamma).$$

Since $1-|h|/(2N)\geqslant\frac{1}{2}$ whenever $|h|\leqslant N$ it follows that

$$2^{-l}\sum_{\boldsymbol{n}\in\mathcal{N}}|T(\boldsymbol{n})|^2\leqslant\sum_{\substack{\boldsymbol{h}\\|h_j|\leqslant 2N_j}}|T(\boldsymbol{h})|^2\prod_{j=1}^{l}(1-|h_j|/(2N_j)).$$

On squaring out and interchanging the order of summation this becomes

$$\sum_{\gamma\in\Gamma}\sum_{\gamma'\in\Gamma}b(\gamma)\bar{b}(\gamma')\prod_{j=1}^{l}\left(\frac{1}{2N_j}\left|\sum_{n=1}^{2N_j}e(n(\gamma_j'-\gamma_j))\right|^2\right).$$

The innermost sum here is

$$\ll\min(N_j,\|\gamma_j'-\gamma_j\|^{-1})\ll N_j/(1+N_j\|\gamma_j'-\gamma_j\|).$$

Therefore

$$\sum_{\boldsymbol{n}\in\mathcal{N}}|T(\boldsymbol{n})|^2\ll\sum_{\gamma\in\Gamma}|b(\gamma)|^2\sum_{\gamma'\in\Gamma}\prod_{j=1}^{l}(N_j(1+N_j\|\gamma_j'-\gamma_j\|)^{-2}).\quad(5.24)$$

Let $0<\delta<1$,

$$F(\beta,\delta)=N\quad\text{or}\quad N(1+N\|\beta\|)^{-2}$$
$$\text{according as }\|\beta\|<\delta\quad\text{or}\quad\|\beta\|\geqslant\delta,$$

and

$$I(\alpha,\delta)=\{\beta:\|\alpha-\beta\|<\delta,\quad 0\leqslant\beta<1\}.$$

Then it suffices to prove that

$$F(\alpha,\delta)\ll\delta^{-1}\int_{I(\alpha,\delta)}F(\beta,\delta)\mathrm{d}\beta,\quad(5.25)$$

for then it follows from (5.24) that

$$\sum_{\boldsymbol{n}\in\mathcal{N}}|T(\boldsymbol{n})|^2\ll\sum_{\gamma\in\Gamma}|b(\gamma)|^2\sum_{\gamma'\in\Gamma}(\delta_1\ldots\delta_l)^{-1}$$
$$\times\prod_{j=1}^{l}\left(\int_{I(\gamma_j'-\gamma_j,\delta_j)}F(\beta,\delta_j)\mathrm{d}\beta\right).$$

By a change of variables the product of the integrals becomes

$$\int_{\mathscr{R}(\gamma')}\prod_{j=1}^{l}F(\beta_j-\gamma_j,\delta_j)\mathrm{d}\boldsymbol{\beta}$$

so that, on the hypothesis concerning $\mathscr{R}(\gamma')$, the sum over γ' is at most

$$(\delta_1\ldots\delta_l)^{-1}\prod_{j=1}^{l}\left(\int_0^1F(\beta-\gamma_j,\delta_j)\mathrm{d}\beta\right).$$

The jth integral here is

$$\ll \int_0^{\delta_j} N \mathrm{d}\beta + \int_{\delta_j}^{\infty} N(1+N\beta)^{-2}\mathrm{d}\beta$$
$$= N\delta_j + (1+N\delta_j)^{-1}.$$

It remains, therefore, to establish (5.25).

If $\|\alpha\| \geqslant \delta$ and $\|\alpha-\beta\| < \delta$, then $1+N\|\beta\| \leqslant 1+N(\|\alpha\|+\delta) \leqslant 2(1+N\|\alpha\|)$, so that $F(\beta,\delta) \geqslant N(1+N\|\beta\|)^{-2} \geqslant \frac{1}{4}F(\alpha,\delta)$ which gives (5.25) when $\|\alpha\| \geqslant \delta$. In the contrary case $\|\alpha\| < \delta$ it can be supposed that $|\alpha| < \delta$, and moreover, if necessary by a change of variable, that $\alpha \geqslant 0$. If $0 \leqslant \beta < \delta$, then $-\delta < \alpha - \beta < \delta$ so that every such β is in $I(\alpha,\delta)$ and $F(\beta,\delta) = N = F(\alpha,\delta)$. Hence

$$\delta^{-1} \int_{I(\alpha,\delta)} F(\beta,\delta)\mathrm{d}\beta \geqslant N = F(\alpha,\delta)$$

which gives (5.25) once more.

The technique adopted here for estimating $f(\boldsymbol{\alpha})$ is to compare

$$\sum_{m\in\mathscr{M}} |g(m,\boldsymbol{\beta})|^{2s}$$

with $J_s(2N)$ by means of Lemma 5.3 and a suitable choice of $\mathscr{M}$.

Let $\boldsymbol{\gamma}(m) = (\gamma_1(m), \ldots, \gamma_{k-1}(m))$ where

$$\gamma_j(m) = \sum_{h=j}^{k} \alpha_h \binom{h}{j} (-m)^{h-j} \quad (1 \leqslant j \leqslant k-1). \tag{5.26}$$

Then, by (5.19),

$$\mathbf{v}^{(k)}(x-m)\cdot\boldsymbol{\alpha} = \mathbf{v}^{(k-1)}(x)\cdot\boldsymbol{\gamma}(m) + x^k\alpha_k + \sum_{j=1}^{k} \alpha_j(-m)^j. \tag{5.27}$$

Thus, in order to apply the lemma to the sum g, given by (5.22), it is necessary to discuss the spacing of the $\gamma_j(m)$ modulo 1.

Suppose that $1 \leqslant x, y \leqslant N$, $x \neq y$ and define

$$\begin{aligned} a_{hj} &= \frac{k!}{h+1}\binom{h+1}{j}\frac{(-x)^{h+1-j}-(-y)^{h+1-j}}{y-x} && (1 \leqslant j \leqslant h < k), \\ a_{hj} &= 0 && (1 \leqslant h < j < k). \end{aligned} \tag{5.28}$$

Note that a_{hj} is an integer and $a_{jj} = k!$. Further define

$$\beta_h = \alpha_{h+1}(h+1)(y-x) \tag{5.29}$$

and $\tau_j = k!(\gamma_j(x) - \gamma_j(y))$. Then, by (5.28),

$$\tau_j = k!(\gamma_j(x) - \gamma_j(y)) = \sum_{h=1}^{k-1} \beta_h a_{hj}. \tag{5.30}$$

The next objective is to invert this linear transformation. Write

$$\boldsymbol{A} = (a_{hj})_{h=1}^{k-1}{}_{j=1}^{k-1}$$

and $\boldsymbol{B} = \boldsymbol{A} - k!\boldsymbol{I}$ where $\boldsymbol{I}$ denotes the unit $(k-1) \times (k-1)$ matrix. The matrix $\boldsymbol{A}$ is lower triangular, and $\boldsymbol{B} = (b_{hj})$ where $b_{hj} = 0$ when $1 \leqslant h \leqslant j < k$ and $b_{hj} = a_{hj}$ when $1 \leqslant j < h < k$. Hence (5.30) can be written in the form

$$\boldsymbol{\tau} = \boldsymbol{\beta}\boldsymbol{A}. \tag{5.31}$$

The tth power of $\boldsymbol{B}$ satisfies

$$\boldsymbol{B}^t = (b_{hj}^{(t)})$$

where

$$b_{hj}^{(t)} = \sum_{j_1=1}^{k-1} \dots \sum_{j_{t-1}=1}^{k-1} b_{hj_1} b_{j_1 j_2} \dots b_{j_{t-1} j}.$$

Therefore $b_{hj}^{(t)}$ is an integer and $b_{hj}^{(t)} = 0$ when $h < j + t$. Moreover, by (5.28), $a_{hj} \ll N^{h-j}$ when $h > j$. Hence, for $h \geqslant j + t$,

$$b_{hj}^{(t)} \ll \underset{j < j_{t-1} < \dots < j_2 < j_1 < h}{\sum_{j_1} \dots \sum_{j_{t-1}}} N^{h-j_1} N^{j_1 - j_2} \dots N^{j_{t-1} - j}$$

So that

$$b_{hj}^{(t)} \ll N^{h-j}. \tag{5.32}$$

Clearly $\boldsymbol{B}^{k-1}$ is the null matrix. Thus, on writing $\boldsymbol{J} = k!\boldsymbol{I}$ and

$$\boldsymbol{D} = \boldsymbol{J}^{k-2} - \boldsymbol{J}^{k-3}\boldsymbol{B} + \dots + (-1)^{k-2}\boldsymbol{B}^{k-2}$$

one obtains

$$\boldsymbol{A}\boldsymbol{D} = (\boldsymbol{B} + \boldsymbol{J})\boldsymbol{D} = \boldsymbol{J}^{k-1} + (-1)^{k-2}\boldsymbol{B}^{k-1} = (k!)^{k-1}\boldsymbol{I}.$$

Hence, by (5.31),

$$\boldsymbol{\tau}\boldsymbol{D} = (k!)^{k-1}\boldsymbol{\beta},$$

so that

$$(k!)^{k-1}\beta_j = (k!)^{k-2}\tau_j + \sum_{t=1}^{k-2} (-1)^t (k!)^{k-2-t} \sum_{h=j+t}^{k-1} \tau_h b_{hj}^{(t)}.$$

Thus, by (5.32),

$$\|(k!)^{k-1}\beta_j\| \ll \sum_{h=j}^{k-1} \|\tau_h\| N^{h-j}.$$

Therefore, by (5.29) and (5.30),

$$\|(k!)^k\alpha_j(x-y)\| \ll \sum_{h=j-1}^{k-1} \|\gamma_j(x)-\gamma_j(y)\| N^{h-j+1} \quad (2\leqslant j\leqslant k). \tag{5.33}$$

Suppose that for some j with $2\leqslant j\leqslant k$ there are a, q with $(a, q)=1$, $q\leqslant N^j$ and $|\alpha_j - a/q|\leqslant q^{-2}$. Let

$$L=\min(q, N). \tag{5.34}$$

Then for each $x\in[1, L]$ the number of $y\in[1, L]$ for which

$$\|(k!)^k\alpha_j(x-y)\| \leqslant N^{i-j}$$

is bounded by the number of $y\in[1, L]$ for which

$$\|(k!)^k a(x-y)/q\| \leqslant N^{1-j}+(k!)^k Lq^{-2}$$

and this is at most R where

$$R=((k!)^k Lq^{-1}+1)(2qN^{1-j}+2(k!)^k Lq^{-1}+1). \tag{5.35}$$

Hence there is a set $\mathscr{M}$ of integers $x\in[1, L]$ such that $M=\operatorname{card}\mathscr{M}$ satisfies $M\geqslant L/(R+1)$ and such that for each pair x, y with $x\in\mathscr{M}$, $y\in\mathscr{M}$, $x\neq y$,

$$\|(k!)^k\alpha_j(x-y)\| > N^{1-j}.$$

By (5.33), for every such pair x, y there exists an h for which $j-1\leqslant h\leqslant k-1$ and

$$\|\gamma_h(x)-\gamma_h(y)\| \gg N^{-h}.$$

Now Lemma 5.3 can be applied with k replaced by $k-1$, with $N_j=sN^j$, with $\delta_j\gg N^{-j}$, with $\Gamma=\{\gamma(m):m\in\mathscr{M}\}$ and with

$$a(n)=\sum_{x_1,\ldots,x_s}{}' \; e((x_1^k+\ldots+x_s^k)\alpha_k+(x_1+\ldots+x_s)\beta)$$

where the sum is restricted to the solutions $x_1,\ldots,x_s$ of the simultaneous equations

$$x_1^h+\ldots+x_s^h=n_h \quad (1\leqslant h\leqslant k-1)$$

with $1\leqslant x_r\leqslant 2N$. Therefore, by (5.22), (5.27) and (5.6),

$$\sum_{m\in\mathscr{M}} |g(m, \beta)|^{2s} \ll J_s^{(k-1)}(2N)N^{k(k-1)/2}.$$

Hence, by (5.23) and Hölder's inequality,

$$f(\boldsymbol{\alpha})^{2s} \ll (R/L)(\log 2N)^{2s} J_s^{(k-1)}(2N)N^{k(k-1)/2}.$$

This with (5.34) and (5.35) gives the following theorem.

Theorem 5.2 *Suppose that there exist* j, a, q *with* $2 \leqslant j \leqslant k$, $|\alpha_j - a/q| \leqslant q^{-2}$, $(a, q) = 1$, $q \leqslant N^j$. *Then*

$$f(\boldsymbol{\alpha}) \ll (J_s^{(k-1)}(2N)N^{k(k-1)/2}(qN^{-j} + N^{-1} + q^{-1}))^{1/2s} \log 2N.$$

Combining this with Theorem 5.1 gives

Theorem 5.3 *On the hypothesis of Theorem* 5.2,

$$f(\boldsymbol{\alpha}) \ll N(N^{\eta}(qN^{-j} + N^{-1} + q^{-1}))^{1/(2(k-1)l)} \log 2N$$

where

$$\eta = \tfrac{1}{2}(k-1)^2\left(\frac{k-2}{k-1}\right)^l.$$

In particular, if $N \ll q \ll N^{j-1}$, *then*

$$f(\boldsymbol{\alpha}) \ll N^{1-\sigma} \log 2N$$

where

$$\sigma = \max_l \frac{1}{2(k-1)l}\left(1 - \tfrac{1}{2}(k-1)^2\left(\frac{k-2}{k-1}\right)^l\right). \tag{5.36}$$

Moreover $4\sigma k^2 \log k \sim 1$ *as* $k \to \infty$.

All but the last part follow at once. To prove the last part observe that when $k \geqslant 3$ the maximum is attained for a value of l satisfying $\left| l - \lambda\left(\log\frac{k-1}{k-2}\right)^{-1}\right| < 1$ where λ is the larger root of the transcendental equation

$$e^{\lambda} = \tfrac{1}{2}(k-1)^2(\lambda + 1).$$

Now it is readily seen that $\lambda \sim 2 \log k$ and

$$\sigma = \frac{1}{2k^2}\frac{1}{\lambda + 1}\left(1 + O\left(\frac{1}{k}\right)\right).$$

A little calculation shows that Theorem 5.3 gives stronger results than Lemma 2.4 when $k \geqslant 12$.

5.3 The minor arcs in Waring's problem

Let $f(\alpha)$ be given by (1.6). Then

$$\int_0^1 |f(\alpha)|^{2s} d\alpha$$

is the number of solutions of

$$x_1^k + \ldots + x_s^k = y_1^k + \ldots + y_s^k$$

with $1 \leqslant x_j, y_j \leqslant N$. Hence

$$\int_0^1 |f(\alpha)|^{2s} d\alpha \ll N^{k(k-1)/2} J_s(N). \tag{5.37}$$

Let $N, P, \mathfrak{M}, \mathscr{U}$ be as in § 4.4 and let $\mathfrak{m} = \mathscr{U} \setminus \mathfrak{M}$. Let $\alpha \in \mathfrak{m}$. Choose a, q so that $(a, q) = 1$, $q \leqslant n/P$ and $|\alpha - a/q| \leqslant Pq^{-1}n^{-1}$ (Lemma 2.1). Then $1 \leqslant a \leqslant q$, and since α lies in no major arc $\mathfrak{M}(q, a)$ it follows that $q > P$. Hence, by Lemma 2.4 and Theorem 5.3

$$f(\alpha) \ll N^{1-\sigma_0+\varepsilon}$$

where $\sigma_0 = \max(\sigma, 2^{1-k})$ and σ is given by (5.36). Thus, by Theorem 5.1 and (5.37) with s replaced by kl, there is a positive number δ such that whenever

$$s > \frac{k^2}{2\sigma_0}\left(1 - \frac{1}{k}\right)^l + 2kl$$

one has

$$\int_{\mathfrak{m}} |f(\alpha)|^{2s} d\alpha \ll N^{2s-k-\delta}.$$

Combined with (4.33) and Theorems 4.4 and 4.6 this yields

Theorem 5.4 *Let $\sigma_0 = \max(\sigma, 2^{1-k})$ and let s_0 denote the least integer such that*

$$s_0 > \min_l \left(\frac{k^2}{2\sigma_0}\left(1 - \frac{1}{k}\right)^l + 2kl \right). \tag{5.38}$$

Then the asymptotic formula (2.27) holds whenever $s \geqslant s_0$. Also $s_0 \sim 4k^2 \log k$ as $k \to \infty$.

Again, a modicum of computation shows that $s_0 < 2^k + 1$ when $k \geqslant 11$.

5.4 An upper bound for $G(k)$

The investigation here of Waring's problem has so far concentrated on an asymptotic formula for the number of solutions of

$$x_1^k + \ldots + x_s^k = n.$$

However, Hardy and Littlewood (1925) observed that the required size of s could be reduced by restricting the range of some of the variables x_j. This technique was later greatly exploited by Vinogradov and Davenport.

Vinogradov has shown that $G(k) \leqslant k(\log k)(C + o(1))$ as $k \to \infty$, and over a period of nearly thirty years has reduced the permissible value of C to 2. There is a fairly simple argument that shows that C can be taken to be 3 and which motivates many of the underlying ideas. For the further reduction from 3 to 2, see Chapter 7.

Let Z be large and write

$$Z_1 = \tfrac{1}{6}Z, \; Z_{j+1} = \tfrac{1}{2}Z_j^{1-1/k},$$

and let $Q_Z(m)$ denote the number of solutions of

$$x_1^k + \ldots + x_t^k = m$$

with $Z_j < x_j \leqslant 2Z_j$. Then $\Sigma_m Q_Z(m)^2$ is the number of solutions of

$$x_1^k + \ldots + x_t^k = y_1^k + \ldots + y_t^k \text{ with } Z_j < x_j, y_j \leqslant 2Z_j. \tag{5.39}$$

Since

$$|x_1^k - y_1^y| \geqslant |x_1 - y_1| k Z_1^{k-1}$$

and

$$|x_2^k + \ldots + x_t^k - y_2^k - \ldots - y_t^k| < 2^k Z_2^k + O(Z_3^k) < kZ_1^{k-1}$$

it follows that (5.39) can only have a solution with $x_1 = y_1$. By repeating this argument it follows that $x_2 = y_2$, $x_3 = y_3$, and so on. Thus

$$\sum_m Q_Z(m)^2 \ll Z_1 \ldots Z_t \ll \left(\sum_m Q_Z(m) \right)^2 (Z_1 \ldots Z_t)^{-1}.$$

Moreover $Z_1 \ldots Z_t \gg Z^{k-k(1-1/k)^t}$ and $Q_Z(m) = 0$ when

$$m > 3^{-k}Z^k + O(Z^{k-1}).$$

Thus

$$\sum_m Q_Z(m)^2 \ll \left(\sum Q_Z(m) \right)^2 Z^{-k+k(1-1/k)^t} \tag{5.40}$$

and

$$Q_Z(m) = 0 \quad \text{when} \quad m > \tfrac{1}{8}Z^k. \tag{5.41}$$

The above argument also shows that $Q_Z(m)$ is 0 or 1, and gives a set $\mathcal{M}$ of natural numbers m not exceeding Z^k for which m is the sum of t kth powers and

$$\text{card}\,\mathscr{M} \gg Z^{k-k(1-1/k)^t}.$$

Thus, for a comparatively small value of t, say $Ck \log k$, the cardinality of $\mathscr{M}$ can be made relatively close to Z^k. This construction, a slight modification of that due to Hardy and Littlewood, is used in two different ways on the minor arcs. Firstly, in an analogous manner to Hua's lemma (Lemma 2.5) in order to save almost N^k, and secondly (and this is Vinogradov's contribution) to save a further small amount on the minor arcs in a similar, but more efficient, way to Weyl's inequality (Lemma 2.4).

Let

$$H(\alpha) = \sum_m Q_N(m)e(\alpha m). \tag{5.42}$$

Then, by Parseval's identity and (5.40),

$$\int_0^1 |H(\alpha)|^2 \mathrm{d}\alpha \ll H(0)^2 N^{-k+k(1-1/k)^t}. \tag{5.43}$$

The following is due essentially to Vinogradov (1947).

Lemma 5.4 *Let*

$$V(\alpha) = \sum_{X/2<p\leqslant X} \sum_{y\leqslant Y} b_y e(\alpha p^k y)$$

where the b_y are arbitrary complex numbers. Suppose that $\alpha = a/q + \beta$ with $|\beta| \leqslant \frac{1}{2}q^{-1}X^{-k}$, $q \leqslant 2X^k$, $(a, q) = 1$, that $Y \gg X^k$, and that when $q \leqslant X$ one has $|\beta| \gg q^{-1}X^{1-k}Y^{-1}$. Then

$$V(\alpha) \ll \left(XY^{1+\varepsilon} \sum_{y\leqslant Y} |b_y|^2\right)^{1/2}.$$

Note that the argument described below can be modified readily so that the interval of summation $(0, Y]$ for y can be replaced by an arbitrary interval of length Y.

Proof By Cauchy's inequality

$$V(\alpha)^2 \ll X \sum_{X/2<p\leqslant X} \left| \sum_{y\leqslant Y} b_y e(\alpha p^k y) \right|^2. \tag{5.44}$$

When $(h, q) = 1$, the number J of solutions of the congruence

$$x^k \equiv h \pmod q$$

satisfies $J \ll q^{\varepsilon}$. Hence there is an $L \ll q^{\varepsilon}$ such that the primes p with $\frac{1}{2}X < p \leqslant X$ can be divided into L classes $\mathscr{P}_1, \ldots, \mathscr{P}_L$ so that for two distinct primes p_1, p_2 in a given class $\mathscr{P}_j$, $p_1^k \equiv p_2^k \pmod{q}$ if and only if $p_1 \equiv p_2 \pmod{q}$.

Consider two such primes p_1, p_2. By the hypothesis

$$\|\alpha(p_1^k - p_2^k)\| \geqslant \|a(p_1^k - p_2^k)/q\| - \tfrac{1}{2}q^{-1}X^{-k}X^k$$
$$\geqslant \tfrac{1}{2}q^{-1}$$

provided that $p_1 \not\equiv p_2 \pmod{q}$. When $q > X$, the elements of $\mathscr{P}_j$ are distinct modulo q. Hence, for $p \in \mathscr{P}_j$, the αp^k are spaced at least $\frac{1}{2}q^{-1}$ apart modulo 1. Therefore, by the one dimensional case of Lemma 5.3 (the large sieve inequality),

$$\sum_{\substack{X/2 < p \leqslant X \\ p \in \mathscr{P}_j}} \left| \sum_{y \leqslant Y} b_y e(\alpha p^k y) \right|^2 \ll Y \sum_{y \leqslant Y} |b_y|^2 \tag{5.45}$$

and the lemma follows easily from (5.44).

When $q \leqslant X$ the argument may be refined as follows. Suppose that $p_1 \equiv p_2 \pmod{q}$ but $p_1 \neq p_2$. Then, by the hypothesis,

$$\|\alpha(p_1^k - p_2^k)\| = \|\beta(p_1^k - p_2^k)\| = |\beta||p_1^k - p_2^k|$$
$$\gg q^{-1}Y^{-1}|p_1 - p_2|.$$

Now $|p_1 - p_2| \geqslant q$, and so, combined with the argument above, this shows that the αp^k are spaced $\gg Y^{-1}$ apart modulo 1. Hence, by Lemma 5.3, one obtains (5.45) once more.

Let $X = N^{1/2}$, $Y = X^k$ and

$$W(\alpha) = \sum_{X/2 < p \leqslant X} \sum_{y} Q_X(y) e(\alpha p^k y).$$

Adopt the notation of §5.3 and suppose that $\alpha \in \mathfrak{m}$. Choose a, q so that $(a, q) = 1$, $q \leqslant 2X^k$, $|\alpha - a/q| \leqslant \frac{1}{2}q^{-1}X^{-k}$. Then $1 \leqslant a \leqslant q$ and since α is not in a major arc, when $q \leqslant N$ one has $|\alpha - a/q| \gg q^{-1}N^{1-k} > q^{-1}X^{1-k}Y^{-1}$. Hence, by Lemma 5.4, (5.40) and (5.41),

$$W(\alpha) \ll W(0)(N^{k(1-1/k)^t - 1 + \varepsilon})^{1/4}.$$

Therefore, by (5.43) and (1.6),

$$\int_{\mathfrak{m}} f(\alpha)^{4k} H(\alpha)^2 W(\alpha) e(-\alpha n) \mathrm{d}\alpha \ll H(0)^2 W(0) n^{3+\varepsilon-\eta}$$

where

$$\eta = \frac{1}{4k} - \frac{5}{4}\left(1 - \frac{1}{k}\right)^t.$$

Thus if t is chosen so that

$$t > (\log 5k) \Big/ \left(- \log \left(1 - \frac{1}{k}\right)\right), \tag{5.46}$$

then it follows that there is a positive number δ such that

$$\int_{\mathfrak{m}} f(\alpha)^{4k} H(\alpha)^2 W(\alpha) e(-\alpha n) \mathrm{d}\alpha \ll H(0)^2 W(0) n^{3-\delta}. \tag{5.47}$$

Now

$$H(\alpha)^2 W(\alpha) = \sum_m Q^*(m) e(\alpha m) \tag{5.48}$$

where

$$Q^*(m) = \sum_{m_1} \sum_{m_2} \sum_{\substack{X/2 < p \leqslant X \\ m_1 + m_2 + p^k y = m}} \sum_y Q_N(m_1) Q_N(m_2) Q_X(y).$$

By (5.41), $Q^*(m) = 0$ when $m > \frac{1}{2}n$. Hence, by Theorems 4.4 and 4.6,

$$\begin{aligned} &\int_{\mathfrak{M}} f(\alpha)^{4k} H(\alpha)^2 W(\alpha) e(-\alpha n) \mathrm{d}\alpha \\ &\quad = \sum_m Q^*(m) \int_{\mathfrak{M}} f(\alpha)^{4k} e(-(n-m)\alpha) \mathrm{d}\alpha \\ &\quad \gg n^3 \sum_m Q^*(m). \end{aligned}$$

Therefore, by (5.47) and (5.48),

$$\int_0^1 f(\alpha)^{4k} H(\alpha)^2 W(\alpha) e(-\alpha n) \mathrm{d}\alpha \gg n^3 H(0)^2 W(0) > 0.$$

On the other hand, the left side is the number of solutions of

$$\begin{aligned} x_1^k + \ldots + x_{4k}^k + y_1^k + \ldots + y_t^k + z_1^k + \ldots \\ + z_t^k + p^k(w_1^k + \ldots + w_t^k) = n \end{aligned}$$

with the x_j, y_j, z_j, w_j, p restricted in various ways. Hence

$$G(k) \leqslant 4k + 3t.$$

The optimal choice of t in (5.46) occurs with $t \sim k \log k$. Thus

$$G(k) \leqslant k(\log k)(3 + o(1)) \quad \text{as} \quad k \to \infty. \tag{5.49}$$

5.5 Exercises

1 Show that if $\mathscr{L}$ is a sequence of natural numbers l_k such that $\sum_{k=1}^{\infty} 1/l_k$ diverges, then for every ε and k_0 there exists a k_1 such that if $A(X)$ is the number of natural numbers n with $n \leqslant X$ which can be written in the form

$$n = \sum_{k_0 < k \leqslant k_1} x_k^{l_k}$$

with the x_k non-negative integers, then $A(x) > X^{1-\varepsilon}$ $(X > X_0(\varepsilon, k_0))$.

2 (Freiman's hypothesis, 1949; Scourfield, 1960). Let $\mathscr{L}$ denote a sequence of natural numbers l_k. Then show that it has the property that to every k_0 there corresponds a k_1 such that every natural number n can be written in the form

$$n = \sum_{k_0 < k \leqslant k_1} x_k^{l_k}$$

with the x_k non-negative integers if and only if $\sum_{k=1}^{\infty} 1/l_k$ diverges.

3 Let s_0 be as in Theorem 5.4. Show that for $2s \geqslant s_0$ one has

$$\sum_{Q < q \leqslant R} \frac{1}{q} \sum_{\substack{a=1 \\ (a,q)=1}}^{q} \left| f\left(\frac{a}{q}\right) \right|^{2s} \ll (N^k Q^{-1} + R) N^{2s-k}.$$

6

Davenport's methods

6.1 Sets of sums of kth powers

It was demonstrated in §5.4 that the upper bound for $G(k)$ could be radically reduced by first of all constructing a set $\mathscr{M}$ of natural numbers m not exceeding Z^k which are the sum of t kth powers. The construction yields card $\mathscr{M} \gg z^{k\alpha}$, where $\alpha = 1 - (1 - 1/k)^t$, and is a slight simplification of one due to Hardy and Littlewood (1925). In fact, in their construction Z_j is as above for $j = 1, \ldots, t - 1$, but they take $Z_t = Z_{t-1}$. The argument proceeds as before until the $(t-1)$th step, when (5.39) reduces to

$$x_{t-1}^k + x_t^k = y_{t-1}^k + y_t^k.$$

For each given pair y_{t-1}, y_t, the number of choices for x_{t-1}, x_t is $\ll Z_t^\varepsilon$. It follows that

$$\begin{aligned}\sum_m Q_Z(m)^2 &\ll Z_1 \ldots Z_{t-1} Z_t^{1+\varepsilon}\\ &\ll \left(\sum_m Q_Z(m)\right)^2 (Z_1 \ldots Z_{t-1} Z_t^{1-\varepsilon})^{-1}.\end{aligned}$$

Moreover

$$Z_1 \ldots Z_t \gg Z^{k - (k-2)(1-1/k)^{t-2}}.$$

Hence, by Cauchy's inequality,

$$\sum_{\substack{m\\ Q_Z(m) > 0}} 1 \gg Z^{k - (k-2)(1-1/k)^{t-2} - \varepsilon}.$$

Let $N_t(X)$ denote the number of natural numbers m not exceeding X such that m is the sum of at most t kth powers. Then this yields

$$N_t(X) > X^{\alpha_t - \varepsilon} (X > X_0(t, \varepsilon)) \tag{6.1}$$

with

$$\alpha_t = 1 - \left(1 - \frac{2}{k}\right)\left(1 - \frac{1}{k}\right)^{t-2}. \tag{6.2}$$

Note that $\alpha_2 = 2/k$.

There have been a number of refinements of this argument, which have been effective in giving improved upper bounds for $G(k)$ when k is relatively small. The following theorem is a generalization of one due to Davenport and Erdős (1939).

Theorem 6.1 *Let* $t \geqslant 3$, $\theta = 1 - 1/k$, $\lambda_1 = 1$,

$$\lambda_2 = \frac{k^2 - \theta^{t-3}}{k^2 + k - k\theta^{t-3}}, \quad \lambda_j = \frac{k^2 - k - 1}{k^2 + k - k\theta^{t-3}}\theta^{j-3} \ (3 \leqslant j \leqslant t),$$

and $Q(m)$ *denote the number of solutions of*

$$x_1^k + \ldots + x_t^k = m \quad \textit{with} \quad Z^{\lambda_j} < x_j < 2Z^{\lambda_j}. \tag{6.3}$$

Then

$$\sum_m Q(m)^2 \ll Z^{\lambda_1 + \ldots + \lambda_t + \varepsilon}.$$

Corollary *The inequality* (6.1) *holds with* $\alpha_t = 1 - \rho$ *where*

$$\rho = \frac{k^3 - 3k^2 + k + 2}{k^3 + k^2 - k^2\theta^{t-3}}\theta^{t-3}. \tag{6.4}$$

The corollary follows by using Cauchy's inequality in the same way as above.

Proof of the theorem Let M_s denote the number of solutions of

$$x_1^k + \ldots + x_s^k = y_1^k + \ldots + y_s^k \tag{6.5}$$

with $Z^{\lambda_j} < x_j, y_j < 2Z^{\lambda_j}$ and $x_s \neq y_s$. Since $M_1 = 0$,

$$\sum_m Q(m)^2 \ll \sum_{s=2}^{t} M_s Z^{\lambda_{s+1} + \ldots + \lambda_t} + Z^{\lambda_1 + \ldots + \lambda_t}. \tag{6.6}$$

Also M_2 is the number of solutions of

$$x_1^k - y_1^k = x_2^x - y_2^k$$

with $x_2 \neq y_2$ and $Z^{\lambda_j} < x_j, y_j < 2Z^{\lambda_j}$. For each given pair x_2, y_2 with $x_2 \neq y_2$ the number of possible choices for x_1, y_1 is $\ll Z^{\varepsilon}$. Hence

$$M_2 \ll Z^{2\lambda_2 + \varepsilon} \ll Z^{\lambda_1 + \lambda_2}. \tag{6.7}$$

For $s \geqslant 3$,

$$M_s = M_s' + 2M_s'' \tag{6.8}$$

where M_s' is the number of solutions of (6.5) with the additional constraint $x_1 = y_1$, and M_s'' is the number with $x_1 > y_1$. Then

$$M_s' \ll Z^{\lambda_1} L_s \tag{6.9}$$

where L_s is the number of solutions of

$$x_2^k + \ldots + x_s^k = y_2^k + \ldots + y_s^k. \tag{6.10}$$

Given $x_2, \ldots, x_s$, the number of $y_2, \ldots, y_s$ is $\ll 1$(c.f. §5.4). Thus

$$L_s \ll Z^{\lambda_2 + \ldots + \lambda_s}. \tag{6.11}$$

Now consider M_s''. The number of choices for x_2, y_2 is $\ll Z^{2\lambda_2}$. For any such choice (6.5) becomes

$$x_1^k - y_1^k + A + \sum_{j=3}^{s} (x_j^k - y_j^k) = 0 \tag{6.12}$$

where A is fixed. Let $h = x_1 - y_1$. Then $x_1^k - y_1^k > hZ^{k-1}$. Also

$$A + \sum_{j=3}^{s} (x_j^k - y_j^k) \ll Z^{k\lambda_2}.$$

Hence $0 < h \ll Z^{k\lambda_2 - k + 1}$, and (6.12) can be rewritten in the form

$$A + (y_1 + h)^k - y_1^k \ll Z^{k\lambda_3}. \tag{6.13}$$

For a given h let y and $y + j$ be two possible values of y_1 for which (6.13) holds. Then

$$(y + j + h)^k - (y + j)^k - (y + h)^k + y^k \ll Z^{k\lambda_3},$$

whence $hjZ^{k-2} \ll Z^{k\lambda_3}$. Thus the number of possible choices for y_1 is

$$\ll 1 + Z^{k\lambda_3 - k + 2} h^{-1}. \tag{6.14}$$

For given x_1, y_1, (6.12) becomes

$$A_1 + \sum_{j=3}^{s} (x_j^k - y_j^k) = 0 \tag{6.15}$$

where A_1 is fixed. The number of choices for $y_3, \ldots, y_{s-1}$ is $\ll Z^{\lambda_3 + \ldots + \lambda_{s-1}}$ and for any such choice the number of choices for $x_3, \ldots, x_{s-1}$ is $\ll 1$ (observe that $x_4^k + \ldots + x_s^k \ll Z^{\lambda_3 \theta}$ and that there are $\ll 1$ values of x_3^k in an interval of length $Z^{\lambda_3 \theta}$, and so on).

Given $y_3, \ldots, y_{s-1}, x_3, \ldots, x_{s-1}$, (6.15) becomes

$$A_2 + x_s^k - y_s^k = 0$$

where A_2 is fixed, and since $x_s \neq y_s$ the number of choices for x_s, y_s is $\ll Z^{\varepsilon}$. Therefore, by (6.14),

$$M_s'' \ll Z^{2\lambda_2} \sum_{0<h\ll Z^{k\lambda_2-k+1}} (1+Z^{k\lambda_3-k+2}h^{-1})Z^{\lambda_3+\dots+\lambda_{s-1}+\varepsilon}.$$

Thus, by (6.8), (6.9), (6.11),

$$M_s \ll Z^{\lambda_1+\dots+\lambda_s} + Z^{2\lambda_2}(Z^{k\lambda_2-k+1}+Z^{k\lambda_3-k+2})Z^{\lambda_3+\dots+\lambda_{s-1}+2\varepsilon}.$$

The theorem now follows from (6.6) on observing that for $s=3,\dots,t$, $(k+1)\lambda_2-k\leqslant\lambda_s$ and $\lambda_2+k\lambda_3-k+1\leqslant\lambda_s$.

The following theorem is due to Davenport (1942*a*).

Theorem 6.2 *Suppose that* $1\leqslant j\leqslant k-2$, $0<v<1$, $\mathscr{A}$ *is a set of natural numbers a with* $a\leqslant Z^{v+k-1}$, $S=\operatorname{card}\mathscr{A}$, $Q(m)$ *is the number of solutions of*

$$x^k+a=m$$

with $Z<x<2Z$ *and* $a\in\mathscr{A}$, *and* $T=\sum_m Q(m)^2$. *Then*

$$T\ll ZS(1+Z^{v+\varepsilon}(Z^{-2}+Z^{-v-j-1}S)^{2^{-j}}).$$

Proof Let Δ_j be as in § 2.2 and write

$$\mathscr{H}_j=\{\boldsymbol{h}:h_j>0;\ h_1<Z^v;\ h_2,\dots,h_j<Z\}.$$

Let $\rho_j(\boldsymbol{h},m)$ denote the number of solutions of

$$\Delta_j(x^k;\boldsymbol{h})+a=m \quad \text{with} \quad Z<x<2Z, a\in\mathscr{A}, \tag{6.16}$$

and put

$$M_j=\sum_{\boldsymbol{h}\in\mathscr{H}_j}\sum_{a\in\mathscr{A}}\rho_j(\boldsymbol{h},a). \tag{6.17}$$

Clearly

$$T\ll ZS+M_1. \tag{6.18}$$

Also, by Cauchy's inequality,

$$M_j^2<Z^{v+j-1}S\sum_{\boldsymbol{h}\in\mathscr{H}_j}\sum_{a\in\mathscr{A}}\rho_j(\boldsymbol{h},a)^2.$$

The double sum is the number of solutions of

$$\Delta_j(x_1^k;\boldsymbol{h})+a_1=\Delta_j(x_2^k;\boldsymbol{h})+a_2=a$$

with $Z<x_1,x_2<2Z, a_1\in\mathscr{A}, a_2\in\mathscr{A}, a\in\mathscr{A}, \boldsymbol{h}\in\mathscr{H}_j$. Since the elements of $\mathscr{A}$ are distinct this is $\ll M_j + M_{j+1}$. Hence

$$M_j \ll Z^{\nu+j-1}S + (Z^{\nu+j-1}SM_{j+1})^{1/2}.$$

Thus, by induction on j,

$$M_1 \ll Z^{\nu+1-2^{1-j}}S + Z^{(\nu+1)(1-2^{-j})-j2^{-j}}S^{1-2^{-j}}M_{j+1}^{2^{-j}}. \quad (6.19)$$

By (6.17), M_{j+1} is the number of solutions of

$$\Delta_{j+1}(x^k; \boldsymbol{h}) + a_1 = a$$

with $Z < x < 2Z$, $\boldsymbol{h} \in \mathscr{H}_{j+1}$, $a_1 \in \mathscr{A}$, $a \in \mathscr{A}$. By Exercise 2.1, when $j \leqslant k-2$, for each pair a_1, a the number of choices for x, $\boldsymbol{h}$ is $\ll Z^\varepsilon$. Thus $M_{j+1} < S^2 Z^\varepsilon$. The theorem now follows from (6.18) and (6.19)

Theorem 6.2 is usually applied iteratively to give lower bounds for $N_t(X)$ for successive values of t. More generally, let $\mathscr{A}$ denote a strictly increasing sequence of natural numbers a with the property that

$$A(X) = \text{card}\,\{a : a \in \mathscr{A}, a \leqslant X\} \quad (6.20)$$

satisfies

$$A(X) > X^{\alpha - \varepsilon} \quad (X > X_0(\varepsilon)), \quad (6.21)$$

where $0 < \alpha < 1$, and let $N(\mathscr{A}, X)$ denote the number of different numbers of the form $x^k + a$ with $x^k + a \leqslant X$ and $a \in \mathscr{A}$. Let $Z = \frac{1}{4}X^{1/k}$. Then, in the notation of Theorem 6.2,

$$N(\mathscr{A}, X) \geqslant \sum_{\substack{m \\ Q(m) \geqslant 0}} 1,$$

and by Cauchy's inequality

$$\Bigg(\sum_{\substack{m \\ Q(m) > 0}} 1\Bigg) \sum_m Q(m)^2 \geqslant \Bigg(\sum_m Q(m)\Bigg)^2 \gg Z^2 S^2$$

where $S = A(Z^{\nu+k-1})$. Hence, by Theorem 6.2,

$$N(\mathscr{A}, X) \gg ZS(1 + Z^{\nu+\varepsilon}(Z^{-2} + Z^{-\nu-j-1}S)^{2^{-j}})^{-1}.$$

Thus, by (6.21),

$$N(\mathscr{A}, X) > X^{\beta - \varepsilon} \quad (X > X_1(\varepsilon)) \quad (6.22)$$

where

$$\beta = \frac{1}{k}(1 + \alpha(k-1) + \tau)$$

and

$$\tau = \max_{1 \leqslant j \leqslant k-2} \sup_{0 < v < 1} (\min(v\alpha, 2^{1-j} - v(1-\alpha),$$
$$(j+1)2^{-j} - (k-1)\alpha 2^{-j} - v(1-\alpha)(1-2^{-j}))).$$

When $j+1 \leqslant (k-1)\alpha$ the above supremum is non-positive, so the maximum occurs for a value of j with $j+1 > (k-1)\alpha$. For such a given value of j the supremum occurs when v is the lesser of the two values given by

$$v\alpha = 2^{1-j} - v(1-\alpha),$$
$$v\alpha = (j+1)2^{-j} - (k-1)\alpha 2^{-j} - v(1-\alpha)(1-2^{-j}),$$

i.e. by

$$v = 2^{1-j}, \quad v = \frac{j+1-(k-1)\alpha}{2^j - 1 + \alpha}.$$

Thus

$$\tau = \alpha \max_{1 \leqslant j \leqslant k-2} \min\left(2^{1-j}, \frac{j+1-(k-1)\alpha}{2^j - 1 + \alpha}\right)$$

Consider the inequality

$$\frac{j+1-(k-1)\alpha}{2^j - 1 + \alpha} \geqslant \frac{j-(k-1)\alpha}{2^{j-1} - 1 + \alpha}.$$

This is equivalent to each of the following inequalities

$$2^{1-j} \geqslant \frac{j+1-(k-1)\alpha}{2^j - 1 + \alpha},$$
$$1 + (k-1)\alpha \geqslant j + 2^{1-j}(1-\alpha). \tag{6.23}$$

The right-hand side of (6.23) is a strictly increasing function of j. Thus if J is the largest value of j such that (6.23) holds, then

$$\tau = \alpha \frac{J+1-(k-1)\alpha}{2^J - 1 + \alpha},$$

and if there is no such value of j, i.e. if $\alpha < 1/k$, then $\tau = \alpha$. This establishes the following theorem.

Theorem 6.3 *Suppose that* $\mathscr{A}$ *satisfies* (6.20) *and* (6.21). *Let* $H = [(k-1)\alpha]$, *and* $J = H + 1$ *when*

$$2^H((k-1)\alpha - H) \geqslant 1 - \alpha \quad \textit{and} \quad H + 1 \leqslant k - 2$$

and $J = H$ *otherwise. Then* $N(\mathscr{A}, X)$ *satisfies* (6.22) *with*

$$\beta = \frac{1}{k}\left(1 + \alpha(k-1) + \alpha\frac{J+1-(k-1)\alpha}{2^J - 1 + \alpha}\right)$$

when $\alpha \geqslant 1/k$ *and* $\beta = 1/k + \alpha$ *when* $\alpha < 1/k$.

It is useful, in the case of fourth powers, to have a slight refinement of this. The above argument is not materially altered if $Q(m)$ is taken to be the number of solutions of $m = x^4 + a$ with $Z < x < 2Z$, $x \equiv r \pmod{16}$, $a \in \mathscr{A}$, $a \leqslant Z^{v+3}$. Likewise the argument that gives (6.1) with (6.2) is essentially unchanged when each x_j is restricted to a given residue class modulo 16. Thus

Theorem 6.4 (Davenport, 1939*c*) *Let* $N_t^{(h)}(X)$ *denote the number of natural numbers n not exceeding X in the residue class h modulo* 16 *which are the sum of t fourth powers. Then for* $t \geqslant 2$ *and* $0 \leqslant h \leqslant \min(t, 16)$,

$$N_t^{(h)}(X) > X^{\alpha_t - \varepsilon} \quad (X > X_0(\varepsilon, t)) \tag{6.24}$$

where

$$\alpha_2 = \frac{1}{2}, \quad \alpha_{t+1} = \frac{3 + 13\alpha_t}{12 + 4\alpha_t}. \tag{6.25}$$

In particular,

$$\alpha_3 = \tfrac{19}{28}, \quad \alpha_4 = \tfrac{331}{412}, \quad \alpha_5 = \tfrac{5539}{6268}. \tag{6.26}$$

Davenport (1942*a*) has given an improvement upon the argument of Theorem 6.2 which is particularly effective when $k = 5$ or 6. With the same assumptions as in Theorem 6.2, let $Q(m)$ denote the number of solutions of

$$x^k + p^k a = m \tag{6.27}$$

with $Z < x < 2Z$, $a \leqslant Z^{v+k-1}$, $\tfrac{1}{2}Z^{1-v} < p^k < Z^{1-v}$, $p \nmid x$. Also, let $Q(m, p)$ denote the number of solutions of (6.27), for a given p, with $Z < x < 2Z$, $a \leqslant Z^{v+k-1}$, $p \nmid x$. Then, by Cauchy's inequality,

$$T = \sum_m Q(m)^2$$

satisfies

$$T \leqslant P \sum_m \sum_p Q(m, p)^2$$

where P is the number of primes p with $\frac{1}{2}Z^{1-\nu} < p^k < Z^{1-\nu}$. For a given prime p and integer r with $p \nmid r$ the number of solutions of $x^k \equiv r \pmod{p^k}$ is 0 or $(k, \phi(p^k))$. Thus the integers x with $p \nmid x$ can be divided into $q(p) = (k, \phi(p^k))$ classes $\mathscr{R}_1, \ldots, \mathscr{R}_{q(p)}$ such that, if x and y are in a given class $\mathscr{R}_r$, then $x^k \equiv y^k \pmod{p^k}$ if and only if $x \equiv y \pmod{p^k}$. Let $Q_r(m, p)$ denote the number of solutions of (6.27) with $Z < x < 2Z$, $a \leqslant Z^{\nu + k - 1}$, $x \in \mathscr{R}_r$. Then, by Cauchy's inequality,

$$T \leqslant P \sum_m \sum_p \left(\sum_{r=1}^{q(p)} Q_r(m, p) \right)^2$$
$$\leqslant kP \sum_{r=1}^{k} \sum_p \sum_m Q_r(m, p)^2$$

where $Q_r(m, p)$ is taken to be 0 when $r > q(p)$. The triple sum here is bounded by the number of solutions of

$$x_1^k + p^k a_1 = x_2^k + p^k a_2$$

with $x_1 \equiv x_2 \pmod{p^k}$ and x_1, x_2, a_1, a_2, p satisfying the same conditions as before.

Let Δ_j be as above and write

$$\mathscr{H}_j = \{\mathbf{h} : h_i > 0;\, h_1 < 2Z^\nu;\, h_2, \ldots, h_j < Z\}.$$

Let $\rho_j(\mathbf{h}, m, p)$ denote the number of solutions of

$$p^{-k}\Delta_j(x^k; p^k h_1, h_2, \ldots, h_j) + a = m$$

and put

$$M_j = \sum_p \sum_{\mathbf{h} \in \mathscr{H}_j} \sum_{a \in \mathscr{A}} \rho_j(\mathbf{h}, a, p).$$

Then, as in the proof of Theorem 6.2,

$$T \ll P(PZS + M_1)$$

and

$$M_j \ll Z^{\nu + j - 1} PS + (Z^{\nu + j - 1} PSM_{j+1})^{1/2}.$$

Thus, if it can be shown that

$$M_{j+1} \ll S^2 Z^\varepsilon, \tag{6.28}$$

then it follows that

$$T \ll P^2 ZS(1 + Z^{\nu + \varepsilon}(Z^{-2} + Z^{-\nu - j - 1}P^{-1}S)^{2^{-j}}) \tag{6.29}$$

and the extra factor of P^{-1} in the innermost bracket gives an improvement over Theorem 6.2.

It is probable that (6.28) holds whenever $j \leqslant k-3$, but this seems rather difficult to prove in general. However, it can be established for certain values of j. Consider the central difference operator ∇_j which can be defined in terms of Δ_j by

$$\nabla_j(f(\alpha); \beta_1, \ldots, \beta_j) = \Delta_j(f(\alpha - \tfrac{1}{2}\beta_1 - \ldots - \tfrac{1}{2}\beta_j); \beta_1, \ldots, \beta_j).$$

Then

$$\begin{aligned}
&\nabla_j(\alpha^k; \beta_1, \ldots, \beta_j)\\
&= \sum_{\theta_1 = \pm 1} \cdots \sum_{\theta_j = \pm 1} \theta_1 \ldots \theta_j(\alpha + \tfrac{1}{2}\theta_1\beta_1 + \ldots + \tfrac{1}{2}\theta_j\beta_j)^k\\
&= \sum_{l_0} \sum_{\substack{l_1 \\ 2 \nmid l_1}} \cdots \sum_{\substack{l_j \\ 2 \nmid l_j}} \frac{k!}{l_0! l_1! \ldots l_j!} 2^{l_0 - k + j} \alpha^{l_0} \beta_1^{l_1} \ldots \beta_j^{l_j} \quad (l_0 + l_1 + \ldots + l_j = k)\\
&= \beta_1 \ldots \beta_j \sum_{l_0} \cdots \sum_{l_j} \frac{k! 2^{l_0 - k + j} \alpha^{l_0} \beta_1^{2l_1} \ldots \beta_j^{2l_j}}{l_0!(2l_1 + 1)! \ldots (2l_j + 1)!} \quad (l_0 + 2(l_1 + \ldots + l_j) = k - j).
\end{aligned}$$

If $k-j$ is odd, then $l_0 \geqslant 1$ in every term, and so

$$\nabla_j(\alpha^k; \beta_1, \ldots, \beta_j) = \alpha\beta_1 \ldots \beta_j p_j(\alpha; \beta_1, \ldots, \beta_j)$$

where

$$\begin{aligned}
&p_j(\alpha; \beta_1, \ldots, \beta_j)\\
&= \sum_{l_0} \cdots \sum_{l_j} \frac{k! 2^{l_0 + 1 - k + j} \alpha^{l_0} \beta_1^{2l_1} \ldots \beta_j^{2l_j}}{(l_0 + 1)!(2l_1 + 1)! \ldots (2l_j + 1)!} \quad (l_0 + 2(l_1 + \ldots + l_j) = k - j - 1).
\end{aligned}$$

If $k - j = 2$, then

$$\nabla_j(\alpha^k; \beta_1, \ldots, \beta_j) = \beta_1 \ldots \beta_j \frac{2^j k!}{2^k 3!}(12\alpha^2 + \beta_1^2 + \ldots + \beta_j^2).$$

The number M_{j+1} can now be reinterpreted as the number of solutions of

$$p^{-k}\nabla_{j+1}(\alpha^k; h_1p^k, h_2, \ldots, h_{j+1}) + a_1 = a_2$$

with $\alpha = x + \frac{1}{2}h_1p^k + \ldots + \frac{1}{2}h_{j+1}$. When $k - j - 1$ is odd and positive

$$\begin{aligned}
p^{-k}&\nabla_{j+1}(\alpha^k; h_1p^k, h_2, \ldots, h_{j+1})\\
&= \alpha h_1 \ldots h_{j+1} p_{j+1}(\alpha; h_1p^k, h_2, \ldots, h_{j+1})
\end{aligned}$$

which is positive. Given a_1, a_2 the number of choices for α, $h_1, \ldots, h_{j+1}$, i.e. for $x, h_1, \ldots, h_{j+1}$, is $\ll Z^{\varepsilon}$. If moreover $k-j-1 \geqslant 3$, then $p_{j+1}(\alpha; \beta_1, \ldots, \beta_{j+1})$ is a polynomial in β_1 of degree at least 2. Thus given $a_1, a_2, \alpha, h_1, \ldots, h_{j+1}$, the number of choices for p is $\ll 1$. Hence in this case one has (6.28).

When $k-j-1=2$,

$$\begin{aligned} & p^{-k}\nabla_{j+1}(\alpha^k; h_1 p^k, h_2, \ldots, h_{j+1}) \\ & = h_1 \ldots h_{j+1} \frac{2^{j+1}k!}{2^k 3!}(12\alpha^2 + p^{2k}h_1^2 + h_2^2 + \ldots + h_{j+1}^2). \end{aligned}$$

Given a_1, a_2, the number of choices for $h_1, \ldots, h_{j+1}$ is $\ll X^{\varepsilon}$. Then given $a_1, a_2, h_1, \ldots, h_{j+1}$, the number of choices for α, p, i.e. for x, p, is again $\ll X^{\varepsilon}$, since the number of solutions of $3u^2 + v^2 = m$ is $\ll m^{\varepsilon}$. Thus, if $j = k-3$, then (6.28) holds once more. This yields

Theorem 6.5 (Davenport, 1942*a*) *Suppose that* $1 \leqslant j \leqslant k-4$ *and* $k-j$ *is even, or that* $j = k-3$. *Suppose further that* $0 < \nu < 1$ *and* $\mathscr{A}$ *is a set of natural numbers a with* $a \leqslant Z^{\nu+k-1}$. *Let* $Q(m)$ *denote the number of solutions of*

$$x^k + p^k a = m$$

with $Z < x < 2Z$, $a \in \mathscr{A}$, $\frac{1}{2}Z^{1-\nu} < p^k < Z^{1-\nu}$, $p \nmid x$, *let* $T = \sum_m Q(m)^2$, *and let* $S = \operatorname{card} \mathscr{A}$. *Then*

$$T \ll P^2 Z S(1 + Z^{\nu+\varepsilon}(Z^{-2} + Z^{-\nu-j-1}P^{-1}S)^{2^{-j}})$$

where P is the number of primes p with $\frac{1}{2}Z^{1-\nu} < p^k < Z^{1-\nu}$.

Corollary *Suppose that* (6.1) *holds, that* $1 \leqslant j \leqslant k-4$ *and* $k-j$ *is even, or that* $j = k-3$. *Then*

$$N_{t+1}(X) > X^{\alpha_{t+1}-\varepsilon} \quad (X > X_0(t+1, \varepsilon))$$

with

$$\alpha_{t+1} = \frac{1}{k}(1 + \alpha_t(k-1) + \tau_j)$$

and

$$\tau_j = \alpha_t \min\left(2^{1-j}, \frac{j+1-(k-1)\alpha_t + k^{-1}}{2^j - 1 + \alpha_t + k^{-1}}\right).$$

This follows from Theorem 6.5 in the same way that Theorem 6.3 follows from Theorem 6.2.

Suppose that $k=5$. Then (6.2) gives $\alpha_2=\frac{2}{5}$, the above corollary gives

$$\alpha_{t+1}=\frac{16+85\alpha_t}{5(16+5\alpha_t)} \quad \text{when} \quad \tfrac{2}{5}\leqslant\alpha_t<\tfrac{3}{5}$$

and Theorem 6.3 gives

$$\alpha_{t+1}=\frac{7+33\alpha_t}{5(7+\alpha_t)} \quad \text{when} \quad \tfrac{3}{5}\leqslant\alpha_t<1.$$

Hence

Theorem 6.6 (Davenport, 1942*a*) *When* $k=5$, (6.1) *holds with*

$$\alpha_2=\tfrac{2}{5}, \quad \alpha_3=\tfrac{5}{9}, \quad \alpha_4=\tfrac{569}{845}, \quad \alpha_8=\tfrac{6\,913\,439}{7\,576\,115}\,(>0.912\,53).$$

Now suppose that $k=6$. Then (6.12) gives $\alpha_2=\frac{1}{3}$, and the corollary to Theorem 6.5 gives

$$\alpha_{t+1}=\frac{19+120\alpha_t}{6(19+6\alpha_t)} \quad \text{when} \quad \tfrac{1}{3}\leqslant\alpha_t<\tfrac{19}{42}$$

$$\alpha_{t+1}=\frac{43+246\alpha_t}{6(43+6\alpha_t)} \quad \text{when} \quad \tfrac{19}{42}\leqslant\alpha_t<\tfrac{2}{3}$$

and Theorem 6.3 gives

$$\alpha_{t+1}=\frac{15+81\alpha_t}{6(15+\alpha_t)} \quad \text{when} \quad \tfrac{2}{3}\leqslant\alpha_t<1.$$

Hence

Theorem 6.7 (Davenport, 1942 *a*) *When* $k=6$, (6.1) *holds with*

$$\alpha_2=\tfrac{1}{3}, \quad \alpha_3=\tfrac{59}{126}, \quad \alpha_4=\tfrac{1661}{2886}, \quad \alpha_5=\tfrac{5549}{8379}, \quad \alpha_6=\tfrac{575\,117}{787\,182},$$

$$\alpha_{13}=\tfrac{24\,040\,980\,990\,984\,981}{25\,335\,323\,032\,000\,606}\,(>0.948\,91).$$

6.2 G(4) = 16

It is useful to introduce here the generating function

$$h(\alpha) = \sum_{X < x \leqslant 2X} e(\alpha x^k) \tag{6.30}$$

and the corresponding auxiliary functions

$$w(\alpha) = \sum_{X^k < x \leqslant (2X)^k} \frac{1}{k} x^{1/k - 1} e(\alpha x) \tag{6.31}$$

and

$$W(\alpha, q, a) = q^{-1} S(q, a) w(\alpha - a/q) \tag{6.32}$$

where $S(q, a)$ satisfies (4.10). The following lemma is then immediate from Theorem 4.1 on taking $n = [2X]^k$ and $n = [X]^k$.

Lemma 6.1 *Suppose that* $(a, q) = 1$ *and* $\alpha = a/q + \beta$. *Then*

$$h(\alpha) - W(\alpha, q, a) \ll q^{1/2 + \varepsilon}(1 + X^k|\beta|) \tag{6.33}$$

and if moreover $|\beta| \leqslant (2kq)^{-1}(2X)^{1-k}$, *then*

$$h(\alpha) - W(\alpha, q, a) \ll q^{1/2 + \varepsilon}. \tag{6.34}$$

One reason for this kind of choice for $h(\alpha)$ is that it fits more readily into the area of ideas used in § 6.1. Another reason is exemplified by the next lemma which shows that $h(a/q + \beta)$ decays like $\|\beta\|^{-1}$ as $\|\beta\|$ grows, rather than like $\|\beta\|^{-1/k}$ as for $f(a/q + \beta)$ (cf. Lemma 4.6). This is not usually of vital importance, but can often result in a reduction of technical difficulties.

Lemma 6.2 *Suppose that* $|\beta| \leqslant \frac{1}{2}$. *Then*

$$w(\beta) \ll X(1 + X^k|\beta|)^{-1}.$$

This can be shown in the same way as Lemma 2.8. As an immediate consequence of this and Theorem 4.2 one has

Lemma 6.3 *Suppose that* $(q, a) = 1$. *Then*

$$W(a/q + \beta, q, a) \ll X q^{-1/k}(1 + X^k\|\beta\|)^{-1}.$$

The following theorem is due to Davenport (1939*c*) and is still the best that is known for fourth powers. Exercise 2.2 gives $G(4) \geqslant 16$.

Theorem 6.8 *Suppose that $n \not\equiv 0$ or $-1 \pmod{16}$ and n is sufficiently large. Then n is the sum of fourteen fourth powers.*

Corollary $G(4) = 16$.

Proof of Theorem 6.8 Choose h_1, h_2, j so that

$$h_1 + h_2 + j \equiv n \pmod{16},\ 0 \leqslant h_1 \leqslant 4,\ 0 \leqslant h_2 \leqslant 4,\ 1 \leqslant j \leqslant 6.$$

Let

$$\nu = \frac{243}{1567}, \qquad X = \tfrac{1}{2}n^{1/4}, \tag{6.35}$$

and let $\mathscr{A}(h)$ denote the set of natural numbers a such that $a \leqslant X^{3+\nu}$, $a \equiv h \pmod{16}$ and a is the sum of four fourth powers. Further, let

$$V_r(\alpha) = \sum_{a \in \mathscr{A}(h_r)} e(\alpha a).$$

Then, by Theorem 6.4,

$$V_r(0) > X^{\mu - \varepsilon}, \quad \mu = \frac{3972}{1567}. \tag{6.36}$$

By (6.30) (with $k = 4$),

$$\int_0^1 |h(\alpha) V_r(\alpha)|^2 \mathrm{d}\alpha = \sum_m Q(m)^2$$

where $Q(m)$ is the number of solutions of

$$x^4 + a = m$$

with $X < x < 2X$ and $a \in \mathscr{A}(h_r)$. Hence, by Theorem 6.2 with $k = 4$, $j = 2$,

$$\int_0^1 |h(\alpha) V_r(\alpha)|^2 \mathrm{d}\alpha \ll X V_r(0)(1 + X^{\nu+\varepsilon}(X^{-2} + X^{-\nu-3} V_r(0))^{1/4}).$$

Hence, by (6.36) and Cauchy's inequality,

$$\int_0^1 |h(\alpha)^2 V_1(\alpha) V_2(\alpha)| \mathrm{d}\alpha \ll X^2 V_1(0) V_2(0) X^{\varepsilon - \gamma}, \qquad \gamma = \tfrac{5539}{1567}. \tag{6.37}$$

Define the major arcs $\mathfrak{M}(q, a)$ by taking $P = (2X)/(2k) = X/k$ and

$$\mathfrak{M}(q, a) = \{\alpha : |\alpha - a/q| \leqslant Pq^{-1}n^{-1}\}.$$

Let $\mathfrak{M}$ denote the union of all the $\mathfrak{M}(q, a)$ with $1 \leqslant a \leqslant q \leqslant P$ and $(a, q) = 1$. Then the $\mathfrak{M}(q, a)$ are disjoint and lie in $\mathscr{U} = (Pn^{-1}, 1 + Pn^{-1}]$.

Let $\mathfrak{m} = \mathscr{U} \backslash \mathfrak{M}$. By Weyl's inequality (Lemma 2.4) and the argument used to prove Theorem 2.1,

$$h(\alpha) \ll X^{7/8 + \varepsilon} \qquad (\alpha \in \mathfrak{m}).$$

Hence, by (6.35) and (6.37),

$$\int_{\mathfrak{m}} |h(\alpha)^6 V_1(\alpha) V_2(\alpha)| \mathrm{d}\alpha \ll n^{1/2 - \delta} V_1(0) V_2(0) \tag{6.38}$$

where δ is a suitable positive constant.

In a similar manner to the proof of Theorem 4.4, one has, for $1 \leqslant m \leqslant n$,

$$\int_{\mathfrak{M}} h(\alpha)^6 e(-\alpha m) \mathrm{d}\alpha = I(m) \mathfrak{S}(m) + O(n^{1/2 - \delta}) \tag{6.39}$$

where

$$I(m) = \sum_{\substack{X^4 < x_1 \leqslant (2X)^4 \\ x_1 + \dots + x_6 = m}} \dots \sum_{X^4 < x_6 \leqslant (2X)^4} 4^{-6}(x_1 \dots x_6)^{-3/4}$$

and $\mathfrak{S}(m)$ is the singular series defined in Theorem 4.3. It is readily verified, for instance by considering the $x_1, \dots, x_5$ with $X^4 < x_j \leqslant 2X^4$, that

$$I(m) \gg n^{1/2} \quad \text{when} \quad \tfrac{3}{4}n < m \leqslant n. \tag{6.40}$$

By Lemma 2.15, when $s = 6$ and $p > 2$, one has $M_m^*(p^\gamma) > 0$. Moreover, when $s = 6$, $p = 2$ and $m \equiv j \pmod{16}$ with $1 \leqslant j \leqslant 6$, it is trivial from the definition of M_m^* in § 2.6 that $M_m^*(2^\gamma) > 0$. Hence, by Theorem 4.5,

$$\mathfrak{S}(m) \gg 1 \quad \text{when} \quad m \equiv j \pmod{16}.$$

If $m = n - a_1 - a_2$ with $a_r \in \mathscr{A}(h_r)$, then m satisfies $\tfrac{3}{4}n < m \leqslant n$ and $m \equiv j \pmod{16}$. Hence, by (6.39) and (6.40),

$$\int_{\mathfrak{M}} h(\alpha)^6 V_1(\alpha) V_2(\alpha) e(-\alpha n) \mathrm{d}\alpha = J(n) + O(n^{1/2 - \delta} V_1(0) V_2(0))$$

where $J(n) \gg n^{1/2}V_1(0)V_2(0)$. Hence, by (6.38)

$$R(n) = \int_0^1 h(\alpha)^6 V_1(\alpha)V_2(\alpha)e(-\alpha n)\mathrm{d}\alpha$$

satisfies

$$R(n) \gg n^{1/2}V_1(0)V_2(0) > 0.$$

Hence n is the sum of fourteen fourth powers as required.

6.3 Davenport's bounds for $G(5)$ and $G(6)$

Theorem 6.9 (Davenport, 1942*b*) $G(5) \leqslant 23$, $G(6) \leqslant 36$.

The proof of this is similar to but simpler than that of Theorem 6.8. On this occasion it suffices to adopt the notation of § 4.4, so that (4.29), . . . , (4.32) hold.

Let $r = 7$, $t = 8$ when $k = 5$, and $r = 10$, $t = 13$ when $k = 6$. Further, let $\mathscr{A}$ denote the set of natural numbers a not exceeding $\frac{1}{8}n$ for which a is the sum of t kth powers and write

$$V(\alpha) = \sum_{a \in \mathscr{A}} e(\alpha a).$$

By Theorems 6.6 and 6.7,

$$\int_0^1 |V(\alpha)|^2 \mathrm{d}\alpha < V(0)^2 n^{-\mu}$$

where $\mu = 0.912\,53$ when $k = 5$, and $\mu = 0.948\,91$ when $k = 6$.

Let $\mathfrak{m} = \mathscr{U} \setminus \mathfrak{M}$. Then, by Weyl's inequality (Lemma 2.4),

$$\int_{\mathfrak{m}} |f(\alpha)^r V(\alpha)^2| \mathrm{d}\alpha \ll n^{r/k-1-\delta} V(0)^2 \tag{6.41}$$

where δ is a suitable fixed positive number.

By Theorem 4.4, when $1 \leqslant m \leqslant n$,

$$\int_{\mathfrak{M}} f(\alpha)^r e(-\alpha m)\mathrm{d}\alpha = Cm^{r/k-1}\mathfrak{S}(m) + O(n^{r/k-1-\delta}) \tag{6.42}$$

where C is a positive number depending only on k and r.

By Lemma 2.15 with $s = r$, $k = 5$ or 6 and n replaced by m, one has $M_m^*(p^\gamma) > 0$. Hence, by Theorem 4.5, $\mathfrak{S}(m) \gg 1$. It now follows easily from (6.41) and (6.42) that

$$\int_0^1 f(\alpha)^r V(\alpha)^2 e(-\alpha n)\mathrm{d}\alpha \gg n^{r/k-1} V(0)^2 > 0,$$

and therefore that $G(k) \leqslant r + 2t$ when $k = 5$ or 6.

6.4 Exercises

1 Show that for $X > X_0$ one has

$$N_{19}(X) > X^{0.9668} \quad \text{when } k = 7,$$
$$N_{28}(X) > X^{0.9838} \quad \text{when } k = 8.$$

Deduce that $G(7) \leqslant 53^\dagger$, $G(8) \leqslant 73$.

2 (Davenport, 1939*a*) Let $Q(m)$ denote the number of solutions of $m = x^3 + y^3 + z^3$ with $Z < x \leqslant 2Z$, $Z^{4/5} < y \leqslant 2Z^{4/5}$, $Z^{4/5} < z \leqslant 2Z^{4/5}$. Show that

$$\sum_m Q(m)^2 \ll Z^{13/5+\varepsilon}.$$

Deduce that (i) $G(3) \leqslant 8$, and (ii) almost every natural number is the sum of four positive cubes.

3 (Davenport, 1950) Show that when $k = 3$ one has

$$N_3(X) > X^{47/54-\varepsilon} \quad (X > X_0(\varepsilon)).$$

† Note that the claim $G(7) \leqslant 52$ of Sambasiva Rao (1941) is based on an arithmetical error.

7

Vinogradov's upper bound for $G(k)$

7.1 Some remarks on Vinogradov's mean value theorem

For the purposes of this chapter, the notation of Chapter 5 is assumed.

By (5.3), $J_s^{(k)}(X, 0, \boldsymbol{h})$ is the number of solutions of

$$\sum_{r=1}^{s} (x_r^j - y_r^j) = h_j \quad (1 \leqslant j \leqslant k) \quad \text{with} \quad 0 < x_r, y_r \leqslant X. \tag{7.1}$$

This system of equations is not soluble when $|h_j| \geqslant sX^j$ for some j. Hence, by (5.4),

$$\sum_{\boldsymbol{h}} J_s^{(k)}(X, 0, \boldsymbol{h}) \ll X^{k(k+1)/2} J_s(X). \tag{7.2}$$

On the other hand, the left side of (7.2) counts all the solutions of (7.1) with $\boldsymbol{h}$ considered as an additional variable. Thus

$$J_s(X) \gg X^{2s - k(k+1)/2}.$$

Recall that $J_s(X)$ is the number of solutions of

$$\sum_{r=1}^{s} (x_r^j - y_r^j) = 0 \quad (1 \leqslant j \leqslant k) \quad \text{with} \quad 0 < x_r, y_r \leqslant X. \tag{7.3}$$

Obviously the number $T_s(x)$ of 'trivial' solutions obtained by taking the y_r to be a permutation of the x_r satisfies

$$[X]^s \leqslant T_s(X) \leqslant s!X^s.$$

Thus

$$J_s(X) \gg \max(X^{2s - k(k+1)/2}, X^s) \tag{7.4}$$

which shows, incidentally, that (7.3) has 'non-trivial' solutions whenever $s > \frac{1}{2}k(k+1)$ and X is sufficiently large. For further comments see § 29 of Hua (1959).

It can be conjectured that, when $k \geqslant 3$, as $X \to \infty$

$$J_s(X) \sim C_{s,k} \max(X^{2s - k(k+1)/2}, X^s). \tag{7.5}$$

While this probably lies very deep, at any rate it is possible to establish it when s is sufficiently large. This is done by adapting the Hardy–

Littlewood method to the k-dimensional unit hypercube $\mathscr{U}_k$. The minor arcs are dealt with by applying Theorems 5.1 and 5.3 which can be thought of as the analogues of Hua's lemma and Weyl's inequality respectively. For the major arcs it is necessary to develop an asymptotic approximation for the generating function

$$f(\boldsymbol{\alpha}) = \sum_{x \leqslant X} e(\alpha_1 x + \ldots + \alpha_k x^k) \tag{7.6}$$

and to estimate the corresponding auxiliary functions

$$I(\boldsymbol{\beta}) = \int_0^X e(\beta_1 \gamma + \ldots + \beta_k \gamma^k) d\gamma, \tag{7.7}$$

$$S(q, \mathbf{a}) = S(q, a_1, \ldots, a_k) = \sum_{x=1}^{q} e((a_1 x + \ldots + a_k x^k)/q). \tag{7.8}$$

7.2 Preliminary estimates

Much of the material in this section is due to Hua (1940*a*, 1952, 1965).

It is convenient here to recall that the polynomial congruence

$$\phi(x) = b_0 + b_1 x + \ldots + b_k x^k \equiv 0 \pmod p$$

is said to have a root of multiplicity m at x_0 when $\phi(x) = (x - x_0)^m \phi_1(x) + p\phi_2(x)$ with $\phi_1(x)$ and $\phi_2(x)$ polynomials such that $p \nmid \phi_1(x_0)$.

Theorem 7.1 *Suppose that* $(q, a_1, \ldots, a_k) = 1$. *Then*

$$S(q, \mathbf{a}) \ll q^{1 - 1/k + \varepsilon}.$$

Proof In a similar manner to the proof of Lemma 2.10, when $(q, r) = (qr, a_1, \ldots, a_k) = 1$ one has

$$S(qr, a_1, \ldots, a_k) = S(q, a_1, ra_2, \ldots, r^{k-1} a_k) \times S(r, a_1, qa_2, \ldots, q^{k-1} a_k).$$

Thus it suffices to treat the case when q is a power of a prime. Suppose that $p \nmid (a_1, a_2, \ldots, a_k)$ and p^τ is the highest power of p dividing $(a_1, 2a_2, \ldots, ka_k)$. Let $x_1, \ldots, x_r$ denote the distinct roots of the congruence

$$p^{-\tau}(a_1 + 2a_2 x + \ldots + ka_k x^{k-1}) \equiv 0 \pmod p$$

and suppose that their respective multiplicities are $m_1, \ldots, m_r$. Note that $r \leqslant k-1$. Further let $m = m_1 + \ldots + m_r$. Then it suffices to show that for $l = 1, 2, \ldots$

$$|S(p^l, a_1, \ldots, a_k)| \leqslant k^2 \max(1, m) p^{l - l/k}. \tag{7.9}$$

Since $m \leqslant k-1$ this easily gives the theorem.

The case $l = 1$. The argument in this case is due to Mordell (1932) and gives more, namely that

$$|S(p, a_1, \ldots, a_k)| \leqslant k p^{1 - 1/k}. \tag{7.10}$$

It can be assumed without loss of generality that $p \nmid a_k$ and $p > k$. Consider

$$T = \sum_{z_1 = 1}^{p} \ldots \sum_{z_k = 1}^{p} |S(p, z_1, \ldots, z_k)|^{2k}. \tag{7.11}$$

Then, by multiplying out the summand, applying (7.8) and inverting the order of summation, one obtains

$$T = p^k M \tag{7.12}$$

where M is the number of solutions of the simultaneous congruences

$$x_1^j + \ldots + x_k^j \equiv y_1^j + \ldots + y_k^j \pmod p \qquad (1 \leqslant j \leqslant k) \tag{7.13}$$

with $1 \leqslant x_i \leqslant p$, $1 \leqslant y_i \leqslant p$. In a similar manner to the proof of Lemma 5.1, it follows that if $x_1, \ldots, x_k, y_1, \ldots, y_k$ satisfy (7.13), then, for each x, $\prod_i (x - x_i) \equiv \prod_i (x - y_i) \pmod p$. Thus the $x_1, \ldots, x_k$ are a permutation of the $y_1, \ldots, y_k$. Therefore $M \leqslant k! p^k$, and so, by (7.12),

$$T \leqslant k! p^{2k}. \tag{7.14}$$

When $p \nmid u$, $ux + v$ runs through a complete set of residues modulo p as x does. Let

$$b_j = b_j(u, v) = \sum_{i=j}^{k} a_i \binom{i}{j} u^j v^{i-j}.$$

Then

$$|S(p, a_1, \ldots, a_k)| = |S(p, b_1, \ldots, b_k)|. \tag{7.15}$$

Moreover $b_k = a_k u^k$ and $b_{k-1} = u^{k-1}(vka_k + a_{k-1})$. Thus, as u varies, b_k takes on $(p-1)(k, p-1)^{-1}$ distinct values modulo p, and for a given u, as v varies, b_{k-1} takes on p distinct values modulo p. Hence, by (7.15) and (7.11),

$$\frac{p(p-1)}{(k, p-1)} |S(p, a_1, \ldots, a_k)|^{2k} \leqslant T.$$

Therefore, by (7.14),

$$|S(p, a_1, \ldots, a_k)|^{2k} \leqslant k!2kp^{2k-2} \leqslant k^{2k}p^{2k-2},$$

which gives (7.10) as required.

The case $l > 1$. This is proved by induction on l. Obviously $p^\tau \leqslant k$. Thus, when $2 \leqslant l \leqslant 2\tau + 1$, (7.9) is trivial. Hence it can be assumed that $l \geqslant 2\tau + 2$.

For brevity write $\phi(x) = a_1x + \ldots + a_kx^k$. Recalling that $x_1, \ldots, x_r$ are the distinct solutions of $p^{-\tau}\phi'(x) \equiv 0 \pmod p$ one obtains

$$S(p^l, a_1, \ldots, a_k) = T_0 + \sum_{j=1}^{r} T_j \tag{7.16}$$

where

$$T_j = \sum_{\substack{x=1 \\ x \equiv x_j \pmod p}}^{p^l} e(\phi(x)p^{-l}) \tag{7.17}$$

and

$$T_0 = \sum_{\substack{y=1 \\ p^{\tau+1} \nmid \phi'(y)}}^{p^{l-\tau-1}} \sum_{z=1}^{p^{\tau+1}} e(\phi(p^{l-\tau-1}z + y)p^{-l}). \tag{7.18}$$

Since $l \geqslant 2\tau + 2$ one has

$$\phi(p^{l-\tau-1}z + y) \equiv \phi(y) + p^{l-\tau-1}z\phi'(y) \pmod{p^l}.$$

Hence the innermost sum in (7.18) is zero. Thus it remains to estimate the contribution to (7.16) from the T_j with $j \neq 0$.

When $r = 0$, i.e. $m = 0$, there is nothing to prove. Suppose that $m \geqslant 1$. When $l \leqslant k$, the trivial estimate $|T_j| \leqslant p^{l-1}$ in (7.16) gives

$$|S(p^l, a_1, \ldots, a_k)| \leqslant kp^{l-1} \leqslant kp^{l-l/k}.$$

Thus it can be supposed that $l > k$.

Consider the polynomial in x,

$$\phi(px + x_j) - \phi(x_j) = b_1x + \ldots + b_kx^k$$

where

$$b_i = p^i \sum_{h=i}^{k} a_h \binom{h}{i} x_j^{h-i}.$$

Let p^ρ denote the highest power of p dividing $(b_1, b_2, \ldots, b_k)$. Clearly $\rho \geqslant 1$. If $\rho > k$, then

$$p \Big| \sum_{h=i}^{k} a_h \binom{h}{i} x_j^{h-i} \quad (1 \leqslant i \leqslant k)$$

and so $p|a_k, p|a_{k-1}, \ldots, p|a_1$ contradicting $(p, a_1, \ldots, a_k) = 1$. Thus

$$\rho \leqslant k < l. \tag{7.19}$$

Let $c_i = b_i p^{-\rho}$ and

$$\psi(x) = p^{-\rho}(\phi(px + x_j) - \phi(x_j)) = c_1 x + \ldots + c_k x^k.$$

Then, by (7.17),

$$|T_j| = p^{\rho - 1}|S(p^{l-\rho}, c_1, \ldots, c_k)|. \tag{7.20}$$

Since $p^{-\tau}\phi'(x) \equiv 0 \pmod p$ has a root of multiplicity m_j at x_j one can write $p^{-\tau}\phi'(x)$ in the form

$$p^{-\tau}\phi'(x) = (x - x_j)^{m_j}\phi_1(x) + p\phi_2(x)$$

with $p \nmid \phi_1(x_j)$ and $\deg \phi_2 < m_j$. Now let p^σ denote the highest power of p dividing $(c_1, 2c_2, \ldots, kc_k)$. Then

$$\begin{aligned} p^{-\sigma}\psi'(x) &= p^{1-\sigma-\rho}\phi'(px + x_j) \\ &= p^{1-\sigma-\rho+\tau}(p^{m_j}x^{m_j}\phi_1(px + x_j) \\ &\quad + p\phi_2(px + x_j)). \end{aligned}$$

The coefficients of this polynomial are all integers and at least one is coprime with p. Since $\deg \phi_2 < m_j$ the coefficient of x^{m_j} is

$$p^{1-\sigma-\rho+\tau+m_j}\phi_1(x_j)$$

so that $\sigma + \rho \leqslant 1 + \tau + m_j$. Hence, if $d > m_j$, then the coefficient of x^d is a multiple of p. Hence

$$p^{-\sigma}\psi'(x) \equiv p^{1-\sigma-\rho+\tau}(p^{m_j}x^{m_j}\phi_1(x_j) + p\phi_2(px + x_j)) \pmod p.$$

Therefore the degree of $p^{-\sigma}\psi'(x)$, modulo p, is at most m_j and so the number of solutions of the congruence

$$p^{-\sigma}\psi'(x) \equiv 0 \pmod p,$$

counting multiple solutions multiply, is at most m_j.

Therefore, on the inductive hypothesis, (7.9), with l replaced by $l - \rho$, a_j by c_j, m by m_j one obtains, via (7.19) and (7.20),

$$|T_j| \leqslant k^2 m_j p^{\rho - 1} p^{(l-\rho)(1-1/k)} \leqslant k^2 m_j p^{l - l/k}.$$

The desired conclusion, (7.9), now follows from (7.16) on summing over all $j \geqslant 1$.

The following theorem gives the asymptotic expansion for f on the major arcs.

Theorem 7.2 *Let* $\alpha_j = a_j/q_j + \beta_j$ $(j = 1, \ldots, k)$ *and suppose that* $q = [q_1, \ldots, q_k]$ *and* $A_j = a_j q q_j^{-1}$. *Then*

$$f(\boldsymbol{\alpha}) = q^{-1} S(q, A) I(\boldsymbol{\beta}) + \Delta$$

where

$$\Delta \ll q(1 + |\beta_1| X + |\beta_2| X^2 + \ldots + |\beta_k| X^k).$$

Proof By Lemma 2.6 with

$$c_x = e((A_1 x + \ldots + A_k x^k) q^{-1})$$

and

$$F(\gamma) = e(\beta_1 \gamma + \ldots + \beta_k \gamma^k)$$

and the observation

$$\sum_{x \leqslant \gamma} c_x = \sum_{y=1}^{q} e((A_1 y + \ldots + A_k y^k) q^{-1}) \sum_{\substack{x \leqslant \gamma \\ x \equiv y \pmod q}} 1$$
$$= \gamma q^{-1} S(q, A) + O(q),$$

one obtains

$$f(\boldsymbol{\alpha}) = q^{-1} S(q, A) \left(F(X) X - \int_0^X F'(\gamma) \gamma \, d\gamma \right) + \Delta$$

where

$$\Delta \ll q \left(1 + \int_0^X |\beta_1 + \ldots + k \beta_k \gamma^{k-1}| d\gamma \right).$$

Integration by parts gives the theorem at once.

Theorem 7.3 *The auxiliary function* $I(\boldsymbol{\beta})$ *satisfies*

$$I(\boldsymbol{\beta}) \ll X(1 + |\beta_1| X + \ldots + |\beta_k| X^k)^{-1/k}.$$

Proof It can be assumed that

$$X = 1,$$

for then the general case follows by a change of variable. It can further be supposed that

$$|\beta_1| + \ldots + |\beta_k| \geqslant 1$$

for otherwise the result is trivial. Let

$$Y_j = (|\beta_j| + \ldots + |\beta_k|)^{1/k},$$
$$p_1(\alpha) = \beta_1 + 2\beta_2 \alpha + \ldots + k \beta_k \alpha^{k-1}$$

and

$$\mathscr{A} = \{\alpha : 0 \leqslant \alpha \leqslant 1, |p_1(\alpha)| \geqslant Y_1\}.$$

Then $\mathscr{A}$ can be dissected into $\ll 1$ intervals on each of which $p_1'(\alpha)$ does not change sign. Let $\mathscr{B}$ be a typical interval of this kind. Then integration by parts gives

$$\int_{\mathscr{B}} e(\beta_1\alpha + \ldots + \beta_k\alpha^k)\mathrm{d}\alpha \ll Y_1^{-1}.$$

Thus it remains to show that

$$\mathscr{C}_1 = \{\alpha : 0 \leqslant \alpha \leqslant 1, |p_1(\alpha)| < Y_1\}$$

satisfies

$$\operatorname{meas}(\mathscr{C}_1) \ll Y_1^{-1}. \tag{7.21}$$

The argument now proceeds by iteratively constructing sets $\mathscr{D}_1, \mathscr{C}_2, \mathscr{D}_2, \mathscr{C}_3, \ldots$ as follows. If $\mathscr{C}_1$ is empty, then there is nothing more to prove. Thus it can be supposed that there is an α_1 such that $0 \leqslant \alpha_1 \leqslant 1$ and $|p_1(\alpha_1)| < Y_1$. Now $|p_1(\alpha_1)| \geqslant |\beta_1| - kY_2^k$ so that if $|\beta_1| > 2kY_2^k$, then $\frac{1}{2}|\beta_1| < (|\beta_1| + Y_2^k)^{1/k} < ((1 + 1/(2k))|\beta_1|)^{1/k}$, whence

$$Y_2^k \ll |\beta_1| \ll 1.$$

Hence in this case (7.21) is trivial. Thus it can be supposed that, for a suitable number C_1 depending at most on k, one has

$$|\beta_1| \leqslant C_1 Y_2^k$$

and $|p_1(\alpha)| < C_1 Y_2$ for every $\alpha \in \mathscr{C}_1$. Therefore it suffices to show that

$$\operatorname{meas}(\mathscr{D}_1) \ll Y_2^{-1}$$

where

$$\mathscr{D}_1 = \{\alpha : 0 \leqslant \alpha \leqslant 1, |\alpha - \alpha_1| > Y_2^{-1}, |p_1(\alpha)| < C_1 Y_2\}.$$

Let

$$p_2(\alpha) = \frac{p_1(\alpha) - p_1(\alpha_1)}{\alpha - \alpha_1}.$$

Then $\mathscr{D}_1 \subset \mathscr{C}_2$ where

$$\mathscr{C}_2 = \{\alpha : 0 \leqslant \alpha \leqslant 1, |p_2(\alpha)| < 2C_1 Y_2^2\}.$$

Proceeding in this way, at the jth step one obtains a constant C_{j-1}, a polynomial $p_j(\alpha)$ of degree $k - j$ and considers

$$\mathscr{C}_j = \{\alpha : 0 \leqslant \alpha \leqslant 1, |p_j(\alpha)| < 2C_{j-1} Y_j^j\}.$$

If $\mathscr{C}_j$ is not empty, then there is an α_j such that $|p_j(\alpha_j)| < 2C_{j-1}Y_j^j$. Defining at each step

$$p_{j+1}(\alpha) = \frac{p_j(\alpha) - p_j(\alpha_j)}{\alpha - \alpha_j}$$

it follows that

$$p_j(\alpha) = \sum_{h=0}^{k-j} \gamma_h^{(j)} \alpha^h$$

with

$$\gamma_h^{(j)} = \sum_{i=h}^{k-j} \gamma_{i+1}^{(j-1)} \alpha_{j-1}^{i-h}$$

and

$$\gamma_h^{(1)} = (h+1)\beta_{h+1}.$$

Thus

$$\gamma_0^{(j)} = j\beta_j + O(|\beta_{j+1}| + \ldots + |\beta_k|)$$

and so it can be supposed that there is a number C_j depending at most on k such that

$$|\beta_j| \leqslant C_j Y_{j+1}^k \tag{7.22}$$

and $|p_j(\alpha)| < C_j Y_{j+1}^j$ for every $\alpha \in \mathscr{C}_j$. Let

$$\mathscr{D}_j = \{\alpha : 0 \leqslant \alpha \leqslant 1, |\alpha - \alpha_j| > Y_{j+1}^{-1}, |p_j(\alpha)| < C_j Y_{j+1}^j\}.$$

Then, by (7.22), one desires to show that $\operatorname{meas}(\mathscr{D}_j) \ll Y_{j+1}^{-1}$.

The process may stop because, for some $j \leqslant k-2$, $\mathscr{C}_j$ is empty or the inequality (7.22) is violated. Otherwise for $j \leqslant k-2$ one has $\mathscr{D}_j \subset \mathscr{C}_{j+1}$ and the process continues until one reaches $\mathscr{D}_{k-1}$. Now

$$\gamma_1^{(k-1)} = k\beta_k.$$

Thus

$$\mathscr{D}_{k-1} \subset \{\alpha : |\gamma_0^{(k-1)} k^{-1} \beta_k^{-1} + \alpha| < C_{k-1} |\beta_k|^{-1/k}\}$$

so that

$$\operatorname{meas}(\mathscr{D}_{k-1}) \ll Y_k^{-1}$$

as required.

7.3 An asymptotic formula for $J_s(X)$

Theorem 7.4 *There are a positive constant C_1 and positive numbers $\delta(k)$ and $C_2(k, s)$ such that whenever $s \geqslant k^2(3\log k + \log\log k + C_1)$ one has*

$$J_s(X) = C_2(k, s) X^{2s - k(k+1)/2} + O(X^{2s - k(k+1)/2 - \delta(k)}).$$

Proof Let X denote a large real number, let

$$\lambda = \frac{1}{2k}, \quad Q_1 = X^{1/2}, \quad Q_j = X^{j-\lambda} \quad (2 \leqslant j \leqslant k) \tag{7.23}$$

and let $\mathscr{U}_k^*$ denote the cartesian product of the intervals $(Q_j^{-1}, 1 + Q_j^{-1}]$. When $q_1 \leqslant \frac{1}{2}X^{1/2}$, $q_j \leqslant X^\lambda$ $(2 \leqslant j \leqslant k)$, and $1 \leqslant a_j \leqslant q_j$ with $(q_j, a_j) = 1$, let $\mathfrak{M}(\boldsymbol{q}, \boldsymbol{a})$ denote the cartesian product of the intervals

$$\{\alpha : |\alpha - a_j/q_j| \leqslant q_j^{-1} Q_j^{-1}\}.$$

The major arcs $\mathfrak{M}(\boldsymbol{q}, \boldsymbol{a})$ are pairwise disjoint and contained in $\mathscr{U}_k^*$. Let $\mathfrak{M}$ denote their union. Then the minor arcs $\mathfrak{m}$ are given by $\mathfrak{m} = \mathscr{U}_k^* \backslash \mathfrak{M}$.

By Lemma 2.1, for each $\boldsymbol{\alpha} \in \mathscr{U}_k^*$, there exist $\boldsymbol{q}, \boldsymbol{a}$ such that $(q_j, a_j) = 1$, $|\alpha_j - a_j/q_j| \leqslant q_j^{-1} Q_j^{-1}$ and $q_j \leqslant Q_j$. Let $\mathfrak{n}$ denote the set of $\boldsymbol{\alpha} \in \mathscr{U}_k^*$ for which in addition $q_j > X^\lambda$ for some j with $2 \leqslant j \leqslant k$. Then, by Theorem 5.3 with $l = [4k \log k]$, it follows that there is a positive constant C_3 such that

$$f(\boldsymbol{\alpha}) \ll X^{1-\rho} \quad (\boldsymbol{\alpha} \in \mathfrak{n}) \quad \text{with} \quad \rho^{-1} = C_3 k^3 \log k. \tag{7.24}$$

Now let $\mathfrak{N}$ denote the set of $\boldsymbol{\alpha} \in \mathscr{U}_k^*$ for which there exist $\boldsymbol{q}, \boldsymbol{a}$ such that $(q_j, a_j) = 1$, $|\alpha_j - a_j/q_j| \leqslant q_j^{-1} Q_j^{-1}$, $q_1 \leqslant Q_1$, $q_j \leqslant X^\lambda$ $(2 \leqslant j \leqslant k)$. Thus $\mathfrak{n} \cup \mathfrak{N} = \mathscr{U}_k^*$ (although $\mathfrak{n} \cap \mathfrak{N}$ may not be empty) and $\mathfrak{M} \subset \mathfrak{N}$. Let $\beta_j = \alpha_j - a_j/q_j$, $q = [q_1, \ldots, q_k]$, $A_j = qa_j/q_j$, so that

$$(q, A_1, \ldots, A_k) = 1. \tag{7.25}$$

By Theorem 7.2 and (7.23),

$$\begin{aligned} &f(\boldsymbol{\alpha}) - q^{-1} S(q, \boldsymbol{A}) I(\boldsymbol{\beta}) \\ &\ll q_1 \ldots q_k (1 + X^{1/2} q_1^{-1} + X^\lambda q_2^{-1} + \ldots + X^\lambda q_k^{-1}) \\ &\ll X^{1-\lambda}. \end{aligned} \tag{7.26}$$

If $\boldsymbol{\alpha} \in \mathfrak{m} \ \mathfrak{n}$, so that $\boldsymbol{\alpha} \notin \mathfrak{M}$, then $q \geqslant q_1 > \frac{1}{2}X^{1/2}$. Hence, by Theorem 7.1 and (7.7),

$$q^{-1} S(q, \boldsymbol{A}) I(\boldsymbol{\beta}) \ll X q^{\varepsilon - 1/k} \ll X^{1-\lambda+\varepsilon}.$$

Therefore (7.24) holds with $\mathfrak{n}$ replaced by $\mathfrak{m}$. Let $m = [C_3] + 1$ and $\eta = \frac{1}{2}k^2(1 - 1/k)^t$. Then, by Theorem 5.1,

$$\int_{\mathfrak{m}} |f(\boldsymbol{\alpha})|^{2tk + 2mk^2} \mathrm{d}\boldsymbol{\alpha} \ll X^{2tk + 2mk^2 - k(k+1)/2 + \eta - 2mk^2 \rho}.$$

Moreover, for $t \geqslant 3k \log k + k \log\log k$ one has

$$\eta - 2mk^2\rho < k^2\left(1 - \frac{1}{k}\right)^t - \frac{1}{k \log k} < 0$$

and so when $s \geqslant tk + mk^2$ there is a positive number $\delta = \delta(k)$ such that

$$\int_{\mathfrak{m}} |f(\alpha)|^{2s} \mathrm{d}\alpha \ll X^{2s - k(k+1)/2 - \delta}.$$

It remains, therefore, to treat the major arcs $\mathfrak{M}$. For $\alpha \in \mathfrak{M}(q, a)$, (7.26) holds. Define $V(\alpha) = V(\alpha, q, a)$ when $\alpha \in \mathfrak{M}(q, a)$, $V(\alpha) = 0$ when $\alpha \in \mathfrak{m}$. Then

$$\int_{\mathfrak{M}} |f(\alpha)^{2s} - V(\alpha)^{2s}| \mathrm{d}\alpha \ll X^{2-\lambda} \int_{\mathfrak{U}_k^*} (|f(\alpha)|^{2s-2} + |V(\alpha)|^{2s-2}) \mathrm{d}\alpha. \tag{7.27}$$

By Theorem 5.1, if $s - 1 \geqslant kl$ with $l \geqslant 3k \log k$, then

$$\int_{\mathfrak{U}_k^*} |f(\alpha)|^{2s-2} \mathrm{d}\alpha \ll X^{2s-2-k(k+1)/2+\eta}$$

with $\eta = \frac{1}{2}k^2(1 - 1/k)^l < 1/(2k) = \lambda$. Hence there is a positive number $\delta = \delta(k)$ such that

$$X^{2-\lambda} \int_{\mathfrak{U}_k^*} |f(\alpha)|^{2s-2} \mathrm{d}\alpha \ll X^{2s-k(k+1)/2-\delta}. \tag{7.28}$$

Let $\alpha \in \mathfrak{M}(q, a)$. Then, by (7.25) and Theorems 7.1 and 7.3,

$$V(\alpha) \ll X q^{\varepsilon - 1/k} (1 + |\beta_1| X + \ldots + |\beta_k| X^k)^{-1/k}.$$

Hence

$$\int_{\mathfrak{U}_k^*} |V(\alpha)|^{2t} \mathrm{d}\alpha \ll X^{2t} W Z$$

where

$$W = \sum_{q_1=1}^{\infty} \cdots \sum_{q_k=1}^{\infty} q_1 \ldots q_k [q_1, \ldots, q_k]^{2t(\varepsilon - 1/k)}$$

and

$$Z = \prod_{j=1}^{k} \int_0^{\infty} (1 + \beta_j X^j)^{-2t/k^2} \mathrm{d}\beta_j.$$

When $t > 2k^2$ one has

$$W \ll \sum_{q_1=1}^{\infty} \cdots \sum_{q_k=1}^{\infty} q_1 \ldots q_k (q_1 \ldots q_k)^{-4} < \infty$$

and

$$Z \ll \prod_{j=1}^{k} X^{-j} = X^{-k(k+1)/2}.$$

Therefore, for $s - 1 > 2k^2$,

$$X^{2-\lambda} \int_{\mathfrak{U}_k^*} |V(\alpha)|^{2s-2} d\alpha \ll X^{2s - k(k+1)/2 - \lambda}.$$

This with (7.27) and (7.28) shows that

$$\int_{\mathfrak{M}} |f(\alpha)|^{2s} d\alpha = \int_{\mathfrak{M}} |V(\alpha)|^{2s} d\alpha + O(X^{2s - k(k+1)/2 - \delta})$$

when s satisfies the hypothesis of the theorem with C_1 suitably chosen.

It follows in a straightforward manner from Theorems 7.1 and 7.3, that

$$\int_{\mathfrak{M}} |V(\alpha)|^{2s} d\alpha = \mathfrak{S} J X^{2s - k(k+1)/2} + O(X^{2s - k(k+1)/2 - \delta})$$

where

$$\mathfrak{S} = \sum_{q_1=1}^{\infty} \dots \sum_{q_k=1}^{\infty} \sum_{\substack{a_1=1 \\ (a_1, q_1)=1}}^{q_1} \dots \sum_{\substack{a_k=1 \\ (a_k, q_k)=1}}^{q_k} |q^{-1} S(q, A_1, \dots, A_k)|^{2s}$$

and

$$J = \int_{\mathbb{R}^k} \left| \int_0^1 e(\beta_1 \alpha + \dots + \beta_k \alpha^k) d\alpha \right|^{2s} d\boldsymbol{\beta}$$

Note that $q^{-1} S(q, A_1, \dots A_k) = (q_1 \dots q_k)^{-1} S(q_1 \dots q_k, a_1, \dots a_k)$. Also $\mathfrak{S} < \infty$, $J < \infty$ and so the theorem follows with $C_2(k, s) = \mathfrak{S} J$. The positivity of $C_2(k, s)$ is a consequence of (7.4).

For a more precise analysis, and several applications, of the above theorem, see Hua (1965).

7.4 Vinogradov's upper bound for $G(k)$

It can now be shown, as an application of Theorem 7.4, that $\limsup_{k \to \infty} \frac{G(k)}{k \log k} \leqslant 2$. In many respects the proof builds on the ideas of § 5.4.

Let n denote a large natural number, and write

$$N = [n^{1/k}].$$

Let K denote a natural number with

$$2K < k$$

and put

$$U_1 = [\tfrac{1}{2}N^{1/2}],\ V_1 = [U_1^{1/2}],\ \eta = \frac{2k-2K-1}{2k-1},\ U_{j+1} = [U_j^\eta],$$
$$V_j = [U_j^{1/2}],\ X = \tfrac{1}{2}N^{1/2}. \tag{7.29}$$

Now let $Q(m)$ denote the number of solutions of the equation

$$(U_1 + x_1)^k + \ldots + (U_l + x_l)^k = m$$

with $x_j \leqslant V_j$, where l is a parameter to be determined suitably in terms of k at a later stage.

Consider

$$W(\alpha) = \sum_{X/2 < p \leqslant X} \sum_m Q(m) e(\alpha p^k m).$$

By Hölder's inequality, for any natural number r,

$$W(\alpha)^{2r} \ll X^{2r-1} \sum_{X/2 < p \leqslant X} \left| \sum_m Q(m) e(\alpha p^k m) \right|^{2r}$$
$$= X^{2r-1} \sum_{X/2 < p \leqslant X} \sum_h Q_1(h) e(\alpha p^k h)$$

where

$$Q_1(h) = \sum_{m_1, \ldots, m_{2r}} Q(m_1) \ldots Q(m_{2r})$$

and the summation is over $m_1, \ldots, m_{2r}$ with

$$m_1 + \ldots + m_r - m_{r+1} - \ldots - m_{2r} = h.$$

Hence, in the notation of §§ 4.4 and 5.3, and by Lemma 5.4,

$$W(\alpha)^{2r} \ll X^{2r-1} \left(X U_1^{k+\varepsilon} \sum_h Q_1(h)^2 \right)^{1/2} \quad (\alpha \in \mathfrak{m}). \tag{7.30}$$

The sum $\sum_h Q_1(h)^2$ is the number of solutions of

$$\sum_{j=1}^{l} L_j(\boldsymbol{x}_j) = 0 \tag{7.31}$$

with $\boldsymbol{x}_j \in [1, V_j]^{4r}$ and

$$L_j(\boldsymbol{y}) = (U_j + y_1)^k + \ldots + (U_j + y_{2r})^k$$
$$- (U_j + y_{2r+1})^k - \ldots - (U_j + y_{4r})^k. \tag{7.32}$$

The estimation of $Q_1(h)$ depends on a lemma, the proof of which requires Theorem 7.4.

Lemma 7.1 *Suppose that $r > CK^2 \log K$ where C is a suitably chosen constant. Then the number R_j of different $\mathbf{y}$ in $[1, V_j]^{4r}$ for which $L_j(\mathbf{y})$ lies in a given interval of length $U_j^{k-K-1/2}$ satisfies*

$$R_j \ll V_j^{4r} U_j^{-K}.$$

Proof For brevity, drop the parameter j. By the binomial theorem,

$$L(\mathbf{y}) = \sum_{i=1}^{k} \binom{k}{i} U^{k-i} M_i(\mathbf{y})$$

where

$$M_i(\mathbf{y}) = y_1^i + \ldots + y_{2r}^i - y_{2r+1}^i - \ldots - y_{4r}^i.$$

Since $\mathbf{y} \in [1, V]^{4r}$ and $V = [U^{1/2}]$ one has

$$\sum_{i=2K+1}^{k} \binom{k}{i} U^{k-i} M_i(\mathbf{y}) \ll U^{k-2K-1} V^{2K+1}$$
$$\ll U^{k-K-1/2}.$$

Hence it suffices to show that the number R^* of different $\mathbf{y}$ in $[1, V]^{4r}$ for which

$$\sum_{i=1}^{2K} \binom{k}{i} U^{k-i} M_i(\mathbf{y})$$

lies in a given interval of length U^{k-2K} satisfies

$$R^* \ll V^{4r-1} U^{1-2K}. \tag{7.33}$$

Consider the number R^{**} of $2K$-tuples of integers $z_1, \ldots, z_{2K}$ with $z_i \ll V^i$ and for which

$$\sum_{i=1}^{2K} U^{k-i} z_i$$

lies in a given interval of length U^{k-2K}. The interval can be written in the form

$$((u-1)U^{k-2K} + v, uU^{k-2K} + v]$$

where u and v are integers with $0 \leqslant v < U^{k-2K}$. Then

$$z_{2K} \equiv u (\mathrm{mod}\, U),\ z_{2K-1} \equiv (u - z_{2K}) U^{-1} (\mathrm{mod}\, U),$$

and so on. Thus z_{2K} is determined modulo U, z_{2K-1} is determined modulo U by z_{2K}, and so on up to z_2. Moreover z_1 is determined uniquely by $z_{2K}, \ldots, z_2$ since $0 \leqslant v < U^{k-2K}$. Hence, on recalling that $V^2 \gg U$, one has

$$\begin{aligned} R^{**} &\ll (V^{2K}U^{-1})(V^{2K-1}U^{-1})\ldots(V^2U^{-1}) \\ &= V^{K(2K+1)-1}U^{1-2K}. \end{aligned} \tag{7.34}$$

By Theorem 7.4, given $z_1, \ldots, z_{2K}$, the number of solutions of

$$\binom{k}{i}M_i(\mathbf{y}) = z_i (1 \leqslant i \leqslant 2K) \quad \text{with} \quad \mathbf{y} \in [1, V]^{4r}$$

is $\ll V^{4r-K(2K+1)}$. This with (7.34) gives (7.33) and so the lemma.

Suppose henceforth that the hypothesis of the lemma holds and let $\mathbf{x}_1, \ldots, \mathbf{x}_l$ be a typical solution of (7.31). By (7.32), for $\mathbf{x}_{j+1} \in [1, V_{j+1}]^{4r}$ one has

$$L_{j+1}(\mathbf{x}_{j+1}) \ll U_{j+1}^{k-1}V_{j+1} \ll U_j^{k-K-1/2}.$$

Hence, by (7.29), $L_1(\mathbf{x}_1)$ lies in an interval of length $\ll U_1^{k-K-1/2}$. Thus, by Lemma 7.1, there are $\ll V_1^{4r}U_1^{-K}$ choices for $\mathbf{x}_1$. Then given $\mathbf{x}_1$, $L_2(\mathbf{x}_2)$ lies in an interval of length $U_2^{k-K-1/2}$, and so on. Hence the total number of choices for $\mathbf{x}_1, \mathbf{x}_2, \ldots, \mathbf{x}_l$ is

$$\ll (V_1 \ldots V_l)^{4r}(U_1 \ldots U_l)^{-K}.$$

By (7.29), $(U_1 \ldots U_l)^K \gg U_1^{(k-1/2)(1-\eta^l)}$. Hence, by (7.29) and (7.30), for $\alpha \in \mathfrak{m}$,

$$W(\alpha) \ll XV_1 \ldots V_l(X^{-1}U^{k+\varepsilon-(k-1/2)(1-\eta^l)})^{1/(4r)} \ll W(0)N^{\varepsilon-\rho}$$

where

$$\rho = \frac{1}{16r} - \frac{1}{8r}(k - \tfrac{1}{2})\eta^l.$$

Now take

$$K = [\tfrac{1}{2}\log k],\ l = 3k,\ r = 1 + [CK^2 \log K]. \tag{7.35}$$

Then, by (7.29),

$$\eta^l = \exp\left(l \log\left(1 - \frac{K}{k - \frac{1}{2}}\right)\right) \ll \exp(-3[\tfrac{1}{2}\log k]) \ll k^{-3/2},$$

where the implicit constants are absolute. Thus, if k is sufficiently large,

$$\rho > \frac{1}{C_1(\log k)^3}, \quad = \sigma$$

say, where C_1 is a suitable constant. Thus, in the notation of §5.4 (but with $W(\alpha)$ as above),

$$\int_{\mathfrak{m}} f(\alpha)^{4k} H(\alpha)^2 W(\alpha) e(-\alpha n) d\alpha$$
$$\ll H(0)^2 W(0) n^{3 + (1 - 1/k)^t - \sigma/k}.$$

Choosing t optimally so that $(1 - 1/k)^t < \sigma/k$ gives

$$t \sim k \log k.$$

Then the contribution from the minor arcs is

$$\ll H(0)^2 W(0) n^{3 - \delta}$$

where $\delta = \delta(k)$ is a suitable positive number.

The major arcs can be treated as in § 5.4. It follows that

$$G(k) \leqslant 2t + 4k + l.$$

This with (7.35) yields

Theorem 7.5 *As $k \to \infty$, $G(k) \leqslant k(\log k)(2 + o(1))$.*

For large k this is the best upper bound for $G(k)$ that is known.

7.5 Exercises

1 Show that, when $s \leqslant k$, $J_s(X) = s!X^s + O(X^{s-1})$.

2 Show that, when $k = 2$, $J_3(X) \ll X^3 \log X$, and that (7.5) is false.

3 Let $G_1(k)$ denote the least s such that almost every natural number is the sum of s kth powers. Show that

$$\limsup_{k \to \infty} \frac{G_1(k)}{k \log k} \leqslant 1.$$

8

A ternary additive problem

8.1 A general conjecture

Suppose that $k_1, k_2, \ldots, k_s$ are s integers satisfying

$$2 \leqslant k_1 \leqslant k_2 \leqslant \ldots \leqslant k_s \quad \text{and} \quad \sum_{j=1}^{s} k_j^{-1} > 1. \tag{8.1}$$

Then the arguments discussed above, particularly in Chapters 2 and 4, suggest that the equation

$$\sum_{j=1}^{s} x_j^{k_j} = n \tag{8.2}$$

has a solution in natural numbers $x_1, \ldots, x_s$ whenever

(i) for each prime p and large k the equation (8.2) is soluble modulo p^k with $p \nmid x_j$ for some j;

(ii) n is sufficiently large.

There has been a great deal of work on questions of this kind, much of it rather inconclusive in nature because the treatment of the minor arcs in the present state of knowledge generally requires $\sum k_j^{-1}$ to be appreciably larger than unity.

The smallest value of s for which (8.1) is satisfied is $s = 3$. Then the only case which has been completely solved is that of $k_1 = k_2 = k_3 = 2$, the classical theorem of Legendre on sums of three squares. However, in all the remaining cases it has been shown that almost all numbers can be represented in the form (8.2). The cases with $k_1 = k_2 = 2$ and with $k_1 = 2, k_2 = k_3 = 3$ are due to Davenport & Heilbronn (1937*a*, *b*), the case $k_1 = 2, k_2 = 3, k_2 = 4$ is due to Roth (1949) and the case $k_1 = 2$, $k_2 = 3$, $k_3 = 5$ is due to Vaughan (1980*a*).

The last case is the hardest, and the remainder of this chapter is taken up with its elucidation. The method can be readily adapted to the other cases.

8.2 Statement of the theorem

Let $E(X)$ denote the number of natural numbers not exceeding X and not being the sum of a square, a cube and a fifth power of natural numbers.

Theorem 8.1 *There is a positive number δ such that $E(X) \ll X^{1-\delta}$.*

In general principle the argument is similar to that of § 3.2. An important feature is that the major arcs can be taken to be longer and more numerous than the presence of the cube and fifth power might suggest. However, a large part of the major arcs is treated, in some respects, more like minor arcs.

Another feature of the argument is that there is some difficulty connected with the convergence of the singular series. This is overcome by replacing the singular series by a finite product.

8.3 Definition of major and minor arcs

Let n denote a large natural number and write

$$P_k = (\tfrac{1}{4}n)^{1/k}.$$

Further, let $R(m) = R(m, n)$ denote the number of representations of m in the form

$$m = x_2^2 + x_3^3 + x_5^5$$

with $P_k < x_k \leqslant 2P_k$, and let

$$I(m) = \sum_{y_2} \sum_{y_3} \sum_{y_5} \tfrac{1}{30} y_2^{-1/2} y_3^{-2/3} y_5^{-4/5} \tag{8.3}$$

where the variables of summation satisfy $P_k^k < y_k \leqslant (2P_k)^k$ and $y_2 + y_3 + y_5 = m$.

Also define

$$S_k = S_k(q, a) = \sum_{r=1}^{q} e(ar^k/q), \tag{8.4}$$

$$A(m, q) = \sum_{\substack{a=1 \\ (a,q)=1}}^{q} q^{-3} S_2 S_3 S_5 e(-am/q) \tag{8.5}$$

and

$$\mathfrak{S}(m, X) = \sum_{q \leqslant X} A(m, q). \tag{8.6}$$

The first part of the proof of Theorem 8.1 is the establishment of

Theorem 8.2 *There is a positive constant δ such that for every sufficiently large n*

$$R(m) = I(m)\mathfrak{S}(m, n^{1/2}) + O(n^{1/30-\delta})$$

for all but $\ll n^{1-\delta}$ *values of m with* $n < m \leqslant 2n$.

Proof Let

$$h_k = h_k(\alpha) = \sum_{P_k < x \leqslant 2P_k} e(\alpha x^k), \tag{8.7}$$

$$\delta = 10^{-5}, \quad P = n^{13/30 + 7\delta}, \quad \mathcal{U} = (P/n, 1 + P/n]. \tag{8.8}$$

Then

$$R(m) = \int_{\mathcal{U}} h_2(\alpha)h_3(\alpha)h_5(\alpha)e(-\alpha m)\mathrm{d}\alpha. \tag{8.9}$$

When $1 \leqslant a \leqslant q \leqslant P$ and $(a, q) = 1$, define the major arc $\mathfrak{M}(q, a)$ by

$$\mathfrak{M}(q, a) = \{\alpha : |\alpha - a/q| \leqslant Pq^{-1}n^{-1}\} \tag{8.10}$$

and take $\mathfrak{M}$ to be the union of all the major arcs. As usual, it is easy to show that the $\mathfrak{M}(q, a)$ are disjoint, and the minor arcs $\mathfrak{m}$ are taken to by $\mathcal{U} \setminus \mathfrak{M}$.

There is an important further subdivision of $\mathfrak{M}$. Let $\mathfrak{M}_1$ denote the subset of $\mathfrak{M}$ formed from those $\mathfrak{M}(q, a)$ with $q > n^{1/12}$, and let

$$\mathfrak{N}(q, a) = \{\alpha : |\alpha - a/q| \leqslant n^{3\delta - 14/15}\}. \tag{8.11}$$

Now define $\mathfrak{M}_2$ to be the union of the $\mathfrak{M}(q, a) \setminus \mathfrak{N}(q, a)$ with $1 \leqslant a \leqslant q \leqslant n^{1/12}$ and $(a, q) = 1$. Then, if one writes

$$\mathfrak{n} = \mathfrak{m} \cup \mathfrak{M}_1 \cup \mathfrak{M}_2 \tag{8.12}$$

and

$$R_1(m) = \int_{\mathfrak{n}} h_2(\alpha)h_3(\alpha)h_5(\alpha)e(-\alpha m)\mathrm{d}\alpha,$$

the aim is to show that

$$\sum_m |R_1(m)|^2 \ll n^{16/15 - 3\delta}$$

and

$$\sum_{q \leqslant n^{1/12}} \sum_{\substack{a=1 \\ (a,q)=1}}^{q} \int_{\mathfrak{N}(q,a)} h_2(\alpha)h_3(\alpha)h_5(\alpha)e(-\alpha m)\mathrm{d}\alpha$$
$$= I(m)\mathfrak{S}(m, n^{1/12}) + O(1). \tag{8.13}$$

The first of these estimates will follow from Parseval's identity if it is shown that

$$\int_{\mathfrak{n}} |h_2(\alpha)h_3(\alpha)h_5(\alpha)|^2 \mathrm{d}\alpha \ll n^{16/15-3\delta}. \tag{8.14}$$

8.4 The treatment of $\mathfrak{n}$

The minor arcs $\mathfrak{m}$ can be treated in a straightforward manner. The integral

$$\int_0^1 |h_2^2 h_5^4| \mathrm{d}\alpha$$

is the number of solutions of the equation

$$u^2 - v^2 + x^5 - y^5 + z^5 - t^5 = 0$$

with $P_2 < u, v \leqslant 2P_2$, $P_5 < x, y, z, t \leqslant 2P_5$. The solutions are of three kinds

(i) $u \neq v$,
(ii) $u = v, x \neq y$,
(iii) $u = v, x = y, z = t$.

Hence the total number of solutions is

$$\ll P_5^{4+\varepsilon} + P_2 P_5^{2+\varepsilon} + P_2 P_5^2.$$

Therefore, by (8.3),

$$\int_0^1 |h_2^2 h_5^4| \mathrm{d}\alpha \ll n^{9/10+\varepsilon}. \tag{8.15}$$

Similarly

$$\int_0^1 |h_3^4| \mathrm{d}\alpha \ll n^{2/3+\varepsilon}. \tag{8.16}$$

By Weyl's inequality (Lemma 2.4), it follows that for each $\alpha \in \mathfrak{m}$

$$h_2(\alpha) \ll n^{1/2+\varepsilon}(P^{-1} + n^{-1/2})^{1/2} \ll n^{\varepsilon}(n/P)^{1/2}.$$

Hence, by (8.16),

$$\int_{\mathfrak{m}} |h_2^2 h_3^4| \mathrm{d}\alpha \ll n^{5/3+3\varepsilon} P^{-1}.$$

Therefore, by Schwarz's inequality, (8.15) and (8.8),

$$\int_{\mathfrak{m}} |h_2^2 h_3^2 h_5^2| \mathrm{d}\alpha \ll n^{16/15-3\delta}. \tag{8.17}$$

Let

$$w_k(\beta) = \sum_x (1/k)x^{1/k-1}e(\beta x) \tag{8.18}$$

where the variable of summation satisfies $P_k^k < x \leqslant (2P_k)^k$, and define

$$W_k = W_k(\alpha, q, a) = q^{-1}S_k(q, a)w_k(\alpha - a/q). \tag{8.19}$$

For $\alpha \in \mathfrak{M}$, define ϕ_k, Δ_k by

$$\phi_k = \phi_k(\alpha) = W_k(\alpha, q, a) \ (\alpha \in \mathfrak{M}(q, a)), \ \Delta_k = \Delta_k(\alpha) = h_k - \phi_k. \tag{8.20}$$

The first step in the treatment of $\mathfrak{M}_1 \cup \mathfrak{M}_2$ is to replace h_2 by ϕ_2. By Theorem 4.1, when $\alpha \in \mathfrak{M}$ one has $\Delta_2(\alpha) \ll P^{1/2+\varepsilon}$. Also, in a similar manner to the proof of (8.16),

$$\int_0^1 |h_3^2 h_5^2| \mathrm{d}\alpha \ll n^{8/15}.$$

Therefore, by (8.8),

$$\int_{\mathfrak{M}} |\Delta_2^2 h_3^2 h_5^2| \mathrm{d}\alpha \ll n. \tag{8.21}$$

The next step is to estimate

$$\int_{\mathfrak{M}_1} |\phi_2^2 h_3^2 h_5^2| \mathrm{d}\alpha.$$

To this end it is first necessary to consider the corresponding integrals with the integrand replaced by $|\phi_2^2 h_5^4|$ and $|\phi_2^2 h_3^4|$.

By (8.19) and (8.20),

$$\int_{\mathfrak{M}} |\phi_2^2 h_5^4| \mathrm{d}\alpha$$

$$\ll \sum_{q \leqslant P} \sum_{\substack{a=1 \\ (a,q)=1}}^{q} q^{-2}|S_2|^2 \int_{-1/2}^{1/2} |w_2(\beta)^2 h_5(\beta + a/q)^4| \mathrm{d}\beta, \tag{8.22}$$

and, by (8.18),

$$|w_2(\beta)|^2 = \sum_h b(h)e(-\beta h)$$

where

$$b(h) = \sum_{x,y} \tfrac{1}{4}(xy)^{-1/2} \tag{8.23}$$

with $x - y = h$, $\frac{1}{4}n = P_2^2 < x, y \leqslant (2P_2)^2 = n$. Moreover, by (8.7),

$$|h_5(\alpha)|^4 = \sum_h c(h)e(\alpha h)$$

where

$$c(h) = \sum_{x, y, z, t} 1$$

with $x^5 - y^5 + z^5 - t^5 = h$ and $P_5 < x, y, z, t \leqslant 2P_5$. Therefore

$$\int_{-1/2}^{1/2} |w_2(\beta)^2 h_5(\beta + a/q)^4| \mathrm{d}\beta = \sum_h b(h)c(h)e(ah/q).$$

Hence, by (8.22),

$$\int_{\mathfrak{M}} |\phi_2^2 h_5^4| \mathrm{d}\alpha \leqslant \sum_h b(h)c(h) \sum_{q \leqslant P} \sum_{\substack{a=1 \\ (a,q)=1}}^{q} q^{-2}|S_2|^2 e(ah/q).$$

It is trivial that $|S_2|$, defined by (8.4), is independent of a and satisfies $|S_2|^2 \ll q$. Therefore, by (3.14), for $h \neq 0$,

$$\sum_{\substack{a=1 \\ (a,q)=1}}^{q} q^{-2}|S_2|^2 e(ah/q) \ll q^{-1} \sum_{d|(q,h)} d.$$

Thus

$$\int_{\mathfrak{M}} |\phi_2^2 h_5^4| \mathrm{d}\alpha \ll b(0)c(0)P + \sum_{h \neq 0} b(h)c(h) \sum_{d|h} \sum_{r \leqslant P/d} \frac{1}{r}.$$

It is immediate from (8.23) that $b(h) \ll 1$. Also, in a similar manner to the proof of (8.16), $c(0) \ll n^{2/5+\varepsilon}$. Moreover $\sum_h c(h) \ll n^{4/5}$. Therefore, by (8.8),

$$\int_{\mathfrak{M}} |\phi_2^2 h_5^4| \mathrm{d}\alpha \ll n^{5/6+8\delta}. \tag{8.24}$$

Various forms of Hölder's inequality are applied in order to estimate

$$\int_{\mathfrak{M}_1} |\phi_2^2 h_3^4| \mathrm{d}\alpha$$

and consequently an estimate is required for

$$\int_{\mathfrak{M}} |\phi_2^4| \mathrm{d}\alpha.$$

By (8.20) and Lemma 6.3,

$$\int_{\mathfrak{M}} |\phi_2^4| \mathrm{d}\alpha \ll \sum_{q \leqslant P} q^{-1} \int_0^{1/2} n^2(1 + n\beta)^{-4} \mathrm{d}\beta,$$

so that

$$\int_{\mathfrak{M}} |\phi_2^4| \mathrm{d}\alpha \ll n^{1+\varepsilon}. \tag{8.25}$$

Therefore, by Hölder's inequality and (8.16),

$$\int_{\mathfrak{M}} |\phi_2^2 \Delta_3 h_3^3| \mathrm{d}\alpha \ll (n^{1+\varepsilon})^{1/4} (n^{2/3+\varepsilon})^{3/4} \sup_{\mathfrak{M}} |\phi_2 \Delta_3|.$$

By Lemmas 6.1 and 6.3, for $\alpha \in \mathfrak{M}(q, a)$ one has

$$\phi_2(\alpha)\Delta_3(\alpha) \ll n^{1/2} q^{\varepsilon}.$$

Hence

$$\int_{\mathfrak{M}} |\phi_2^2 \Delta_3 h_3^3| \mathrm{d}\alpha \ll n^{5/4+2\varepsilon}. \tag{8.26}$$

By Schwarz's inequality, (8.25) and (8.16),

$$\int_{\mathfrak{M}} |\phi_2^2 \phi_3 \Delta_3 h_3^2| \mathrm{d}\alpha \ll (n^{1+\varepsilon})^{1/2} (n^{2/3+\varepsilon})^{1/2} \sup_{\mathfrak{M}} |\phi_3 \Delta_3|,$$

and by Lemmas 6.1 and 6.3, when $\alpha \in \mathfrak{M}(q, a)$,

$$\phi_3(\alpha)\Delta_3(\alpha) \ll n^{1/3} q^{1/6+\varepsilon} \ll n^{1/3} P^{1/6+\varepsilon}.$$

Hence, by (8.8),

$$\int_{\mathfrak{M}} |\phi_2^2 \phi_3 \Delta_3 h_3^2| \mathrm{d}\alpha \ll n^{5/4}. \tag{8.27}$$

By Lemmas 6.1 and 6.3, and (8.8),

$$\int_{\mathfrak{M}} |\phi_2^2 \phi_3^2 \Delta_3^2| \mathrm{d}\alpha \ll \sum_{q \leqslant P} q^{1/3+\varepsilon} \int_0^{1/2} \frac{n^{5/3}}{(1+n\beta)^4} (1+n\beta)^2 \mathrm{d}\beta$$
$$\ll n^{5/4}.$$

Therefore, by (8.26) and (8.27),

$$\int_{\mathfrak{M}_1} |\phi_2^2 h_3^4| \mathrm{d}\alpha \ll n^{5/4+\varepsilon} + \int_{\mathfrak{M}_1} |\phi_2^2 \phi_3^4| \mathrm{d}\alpha. \tag{8.28}$$

Now consider the integral on the right. By (8.20), (8.19), Lemma 6.2 and Theorem 4.2,

$$\int_{\mathfrak{M}_1} |\phi_2^2 \phi_3^4| \mathrm{d}\alpha \ll \sum_{n^{1/12} < q \leqslant P} \sum_{\substack{a=1 \\ (a,q)=1}}^{q} q^{-6} |S_2^2 S_3^4| \int_0^{1/2} \frac{n^{7/3}}{(1+n\beta)^6} \mathrm{d}\beta$$
$$\ll n^{5/4} J$$

where

$$J = \sum_{q \leqslant P} F(q), \quad F(q) = \sum_{\substack{a=1 \\ (a,q)=1}}^{q} q^{-4}|S_3^4|.$$

By Theorem 4.2, $F(q) \ll q^{-1/3}$. Thus $\sum_{h=3}^{\infty} F(p^h) \ll p^{-1}$. Also, by Lemmas 4.3 and 4.4, $|S_3(p^l, a)| \ll p^{l/2}$ when $l = 1$ or 2. Hence $\sum_{h=1}^{2} F(p^h) \ll p^{-1}$. Moreover, by Lemma 4.5, F is a multiplicative function of q. Therefore, there is an absolute constant C such that

$$J \leqslant \prod_{p \leqslant P} (1 + Cp^{-1}).$$

Hence, by (8.28) and elementary prime number theory,

$$\int_{\mathfrak{M}_1} |\phi_2^2 h_3^4| \mathrm{d}\alpha \ll n^{5/4 + \varepsilon}.$$

Therefore, by Schwarz's inequality and (8.24),

$$\int_{\mathfrak{M}_1} |\phi_2^2 h_3^2 h_5^2| \mathrm{d}\alpha \ll n^{16/15 - 3\delta}.$$

Hence, by (8.21),

$$\int_{\mathfrak{M}_1} |h_2^2 h_3^2 h_5^2| \mathrm{d}\alpha \ll n^{16/15 - 3\delta}. \tag{8.29}$$

Now consider $\mathfrak{M}_2$. By Lemma 6.3,

$$\int_{\mathfrak{M}_2} |\phi_2^4| \mathrm{d}\alpha \ll \sum_{q \leqslant P} q^{-1} \int_{n^{3\delta - 14/15}}^{1/2} \frac{n^2}{(1 + n\beta)^4} \mathrm{d}\beta$$
$$\ll n^{4/5 + \varepsilon - 9\delta}.$$

By Hua's lemma (Lemma 2.5),

$$\int_0^1 |h_3^8| \mathrm{d}\alpha \ll n^{5/3 + \varepsilon}, \quad \int_0^1 |h_5^8| \mathrm{d}\alpha \ll n^{1 + \varepsilon}.$$

Hence, by Hölder's inequality,

$$\int_{\mathfrak{M}_2} |\phi_2^2 h_3^2 h_5^2| \mathrm{d}\alpha \ll (n^{4/5 + \varepsilon - 9\delta})^{1/2} (n^{5/3 + \varepsilon})^{1/4} (n^{1 + \varepsilon})^{1/4}$$
$$\ll n^{16/15 - 3\delta}.$$

Therefore, by (8.21),

$$\int_{\mathfrak{M}_2} |h_2^2 h_3^2 h_5^2| \mathrm{d}\alpha \ll n^{16/15 - 3\delta}.$$

This with (8.29) and (8.17) gives (8.14).

8.5 The major arcs $\mathfrak{N}(q, a)$

To complete the proof of Theorem 8.2 it remains to establish (8.13). A simple calculation shows that if $k = 2, 3$ or 5, if $q \leqslant n^{1/12}$, and if $|\beta| \leqslant n^{3\delta - 14/15}$, then

$$q^{1/2+\varepsilon}(1 + n|\beta|) \ll (n/q)^{1/k}(1 + n|\beta|)^{-1}.$$

Hence, by Lemmas 6.1 and 6.3, when $\alpha \in \mathfrak{N}(q, a)$ one has

$$h_k(\alpha),\ W_k(\alpha) \ll (n/q)^{1/k}(1 + n|\alpha - a/q|)^{-1}$$

and

$$h_2(\alpha)h_3(\alpha)h_5(\alpha) - W_2(\alpha)W_3(\alpha)W_5(\alpha)$$
$$\ll (n/q)^{5/6}(1 + n|\alpha - a/q|)^{-1}q^{1/2+\varepsilon}.$$

Thus

$$\sum_{q \leqslant n^{1/12}} \sum_{\substack{a=1 \\ (a,q)=1}}^{q} \int_{\mathfrak{N}(q,a)} |h_2h_3h_5 - W_2W_3W_5| \mathrm{d}\alpha \ll 1.$$

Let $\mathfrak{P}(q, a) = \{\alpha : n^{3\delta - 14/15} < |\alpha - a/q| \leqslant \frac{1}{2}\}$. Then, by Lemma 6.3,

$$\sum_{q \leqslant n^{1/12}} \sum_{\substack{a=1 \\ (a,q)=1}}^{q} \int_{\mathfrak{P}(q,a)} |W_2W_3W_5| \mathrm{d}\alpha \ll 1.$$

Therefore, by (8.19), (8.4), (8.5) and (8.6),

$$\sum_{q \leqslant n^{1/12}} \sum_{\substack{a=1 \\ (a,q)=1}}^{q} \int_{\mathfrak{N}(q,a)} h_2h_3h_5 e(-\alpha m) \mathrm{d}\alpha$$
$$= I_1(m)\mathfrak{S}(m, n^{1/12}) + O(1)$$

where

$$I_1(m) = \int_0^1 w_2(\beta)w_3(\beta)w_5(\beta)e(-\beta m)\mathrm{d}\beta.$$

By (8.18) and (8.3), $I_1(m) = I(m)$, which gives (8.13) as required.

8.6 The singular series

The principal difficulty is that $\sum_{q=1}^{\infty} |A(n, q)|$ apparently diverges. This is resolved by approximating to $\mathfrak{S}(m, n^{1/12})$ by a finite Euler product.

Theorem 8.3 *For all except* $\ll n^{1-\delta}$ *values of* m *with* $n < m \leqslant 2n$ *one has*

$$\mathfrak{S}(m, n^{1/12}) = \prod_{p \leqslant n} \left(\sum_{h=0}^{\infty} A(m, p^h) \right) + O(\exp(-(\log n)^{\delta})). \quad (8.30)$$

It is possible that one could show under similar conditions that the finite product can be replaced by the infinite product, perhaps by a method allied to that of Miech (1968). However there are attendant difficulties which the method described here avoids. For a further discussion of this matter in the case $k_1 = 2, k_2 = k_3 = 3$, see Davenport & Heilbronn (1937*a*).

By (8.4), (8.5) and Theorem 4.2,

$$A(m, 1) = 1, \quad A(m, q) \ll q^{-1/30}. \quad (8.31)$$

Thus each of the series on the right of (8.30) converges absolutely.

For the proof of Theorem 8.3, precise estimates are required for $A(m, p^h)$. To this end the following formulae for $S_k(p^h, a)$, valid when $p \nmid a$, are basic. They are consequences of Lemmas 4.3 and 4.4.

When $p > 2$

$$S_2(p^h, a) = \begin{cases} p^{h/2} & (2|h), \\ \left(\frac{a}{p}\right)_L S_2(p, 1)p^{(h-1)/2} & (2\nmid h), \end{cases} \quad (8.32)$$

when $p > 3$

$$S_3(p^h, a) = \begin{cases} p^{[2h/3]} & (h \not\equiv 1 \pmod 3), \\ 0 & (h \equiv 1 \pmod 3),\ p \equiv 2 \pmod 3), \\ S_3(p, a)p^{2(h-1)/3} & (h \equiv p \equiv 1 \pmod 3)), \end{cases} \quad (8.33)$$

and when $p > 5$

$$S_5(p^h, a) = \begin{cases} p^{[4h/5]} & (h \not\equiv 1 \pmod 5)), \\ 0 & (h \equiv 1 \pmod 5),\ p \not\equiv 1 \pmod 5)), \\ S_5(p, a)p^{4(h-1)/5} & (h \equiv p \equiv 1 \pmod 5)). \end{cases} \quad (8.34)$$

Also, when $k = 3$ or 5 and $p \equiv 1 \pmod k$,

$$S_k(p, a) = \sum_{\chi \in \mathscr{A}} \chi(a)\tau(\bar{\chi}) \quad (8.35)$$

where $\mathscr{A}$ denotes the set of $k-1$ non-principal characters χ modulo p with $\chi^k = \chi_0$. Moreover,

$$|\tau(\chi)| = p^{1/2} \quad \text{and} \quad |S_2(p, 1)| = p^{1/2} (p > 2). \quad (8.36)$$

Lemma 8.1 *Suppose that $h \geqslant 1$ and $p > 5$. Then*

$$A(m, p^h) = 0 \quad \text{when} \quad h > 1 \quad \text{and} \quad p^{h-1} \nmid m, \tag{8.37}$$

$$|A(m, p^h)| \leqslant 8p^{-[(h-1)/30]-1}, \tag{8.38}$$

and

$$A(m, p) = \sum_{\chi \in \mathscr{A}(p)} c(\chi)\chi(m) \tag{8.39}$$

where $\mathscr{A}(p)$ is a collection of non-principal characters modulo p,

$$|c(\chi)| \leqslant p^{-1} \quad \text{and} \quad \operatorname{card} \mathscr{A}(p) \leqslant 8. \tag{8.40}$$

Proof By (8.5), (8.32), (8.33), (8.34) and (8.35),

$$A(m, p^h) = \sum_{\chi \in \mathscr{A}(p^h)} b(\chi) \sum_{a=1}^{p^h} \chi(a)e(-ap^{-h}m) \tag{8.41}$$

where $\mathscr{A}(p^h)$ is a subset of the set of characters modulo p and the $b(\chi)$ are suitable complex numbers. When $h > 1$ and $p^{h-1} \nmid m$ the innermost sum is

$$\sum_{x=1}^{p} \chi(x)e(-xp^{-h}m) \sum_{y=1}^{p^{h-1}} e(-yp^{1-h}m) = 0.$$

This gives (8.37).

The proof of (8.38) is by division into eight different cases.

(i) Suppose that $2|h$, $h \not\equiv 1 \pmod 3$ and $h \not\equiv 1 \pmod 5$. Then $\mathscr{A}(p^h)$ consists solely of the principal character and, by (8.32), (8.34) and (8.35), one has

$$|A(m, p^h)| \leqslant p^{\lambda}$$

where $\lambda = \frac{1}{2}h + [2h/3] + [4h/5] - 2h$. The number λ is an integer and does not exceed $-h/30 \leqslant -[(h-1)/30] - 1/30$. Hence (8.38).

In all the remaining cases (8.41) holds with all the elements of $\mathscr{A}(p^h)$ being non-principal characters modulo p. Thus when $p^h | m$ the innermost sum is automatically 0 and (8.38) follows at once. Also, by (8.37) it can be supposed that either $h = 1$ or $h > 1$, $p^{h-1}|m$ and $p^h \nmid m$. In either case the innermost sum in (8.41) is

$$\sum_{a=1}^{p^h} \chi(a)e\left(\frac{a(-m)}{p\,p^{h-1}}\right) = \bar{\chi}\left(-\frac{m}{p^{h-1}}\right)p^{h-1}\tau(\chi). \tag{8.42}$$

(ii) Suppose that $2 \nmid h$, $h \not\equiv 1 \pmod 3$ and $h \not\equiv 1 \pmod 5$. Then

$\mathscr{A}(p^h)$ consists solely of the quadratic character, and

$$|b(\chi)| = p^{h/2+[2h/3]+[4h/5]-3h}.$$

Hence, by (8.42),

$$|A(m, p^h)| = p^\lambda \quad \text{with} \quad \lambda = \tfrac{1}{2}h + [2h/3] + [4h/5] - 2h - \tfrac{1}{2}.$$

The exponent λ is an integer which does not exceed $-h/30 - \frac{1}{2}$. Thus (8.38).

(iii) Suppose that $2|h$, $h \equiv 1 \pmod 3$ and $h \not\equiv 1 \pmod 5$. When $p \not\equiv 1 \pmod 3$, (8.33) gives $A(m, p^h) = 0$. Hence it may be assumed that $p \equiv 1 \pmod 3$. Then card $\mathscr{A}(p^h) = 2$, and $|A(m, p^h)| \leqslant 2p^\lambda$ with $\lambda = \frac{1}{2}h + 2(h-1)/3 + \frac{1}{2} + [4h/5] - 2h - \frac{1}{2}$. The number λ is an integer not exceeding $-h/30 - \frac{2}{3}$. Therefore (8.38) holds.

(iv) Suppose that $2|h$, $h \not\equiv 1 \pmod 3$ and $h \equiv 1 \pmod 5$. The case $p \not\equiv 1 \pmod 5$ is again trivial, so it may be assumed that $p \equiv 1 \pmod 5$. Then $|A(m, p^h)| \leqslant 4p^\lambda$ with $\lambda = \frac{1}{2}h + [2h/3] + 4(h-1)/5 - 2h$, and the argument is completed as before.

(v) Suppose that $2|h$ and $h \equiv 1 \pmod{15}$. The case $p \not\equiv 1 \pmod{15}$ is trivial, and when $p \equiv 1 \pmod{15}$ one obtains $|A(m, p^h)| \leqslant 8p^{\lambda+1/2}$ with $\lambda = \frac{1}{2}h + 2(h-1)/3 + 4(h-1)/5 - 2h$. This is again an integer which does not exceed $-h/30 - \frac{22}{15} < -[(h-1)/30] - 1$. Hence the exponent $\lambda + \frac{1}{2}$ satisfies $\lambda + \frac{1}{2} \leqslant -[(h-1)/30] - \frac{1}{2}$.

(vi) Suppose that $h \equiv 1 \pmod{10}$ and $h \not\equiv 1 \pmod 3$. When $p \not\equiv 1 \pmod 5$, (8.38) is immediate from (8.34), so it may be assumed that $p \equiv 1 \pmod 5$. Then $|A(m, p^h)| \leqslant 4p^{\lambda+1/2}$ with $\lambda = \frac{1}{2}(h-1) + [2h/3] + 4(h-1)/5 - 2h$. As in the previous case the exponent does not exceed $-[(h-1)/30] - \frac{3}{2}$.

(vii) Suppose that $h \equiv 1 \pmod 6$ and $h \not\equiv 1 \pmod 5$. This can be dealt with in a similar way to case (vi).

(viii) Suppose, finally, that $h \equiv 1 \pmod{30}$. Then $A(m, p^h) = 0$ for $p \not\equiv 1 \pmod{15}$. This leaves the possibility $p \equiv 1 \pmod{15}$. Then $|A(m, p^h)| \leqslant 8p^\lambda$ with $\lambda = \frac{1}{2}h + 2(h-1)/3 + \frac{1}{2} + 4(h-1)/5 + \frac{1}{2} - 2h - \frac{1}{2}$. Again λ is an integer not exceeding $-h/30 - \frac{4}{5} - \frac{1}{6} = -(h-1)/30 - 1$.

The proof of the lemma is now completed by establishing (8.39) with (8.40). When $h = 1$, (8.32), (8.33), (8.34) and (8.35) give (8.41) with $b(\chi) = 0$ unless $p \equiv 1 \pmod{15}$. Thus, when $p \not\equiv 1 \pmod{15}$ one obtains (8.39) with (8.40) trivially satisfied.

When $p \equiv 1 \pmod{15}$, (8.41) holds with $\mathscr{A}(p)$ consisting of the characters χ of the form $\chi = \chi_2\chi_3\chi_5$, where χ_k denotes a non-principal character of order k. Thus all the elements of $\mathscr{A}(p)$ are non-principal and card $\mathscr{A}(p) = 8$. Moreover,

$$b(\chi) = S_2(p, 1)\tau(\bar{\chi}_3)\tau(\bar{\chi}_5)p^{-3}$$

Hence, by (8.41) and (8.42),

$$A(m, p) = \sum_{\chi \in \mathscr{A}(p)} S_2(p, 1)\tau(\bar{\chi}_3)\tau(\bar{\chi}_5)p^{-3}\chi(-1)\tau(\chi)\bar{\chi}(m).$$

If a character χ belongs to $\mathscr{A}(p)$ then so does $\bar{\chi}$.

Also, by (8.36), one has

$$|S_2(p, 1)\tau(\bar{\chi}_3)\tau(\bar{\chi}_5)p^{-3}\chi(-1)\tau(\chi)| = p^{-1}.$$

This gives (8.39) with (8.40), as required.

Let the set $\mathscr{B}$ consist of 1 and those natural numbers q such that if $p|q$, then $p \leqslant n$ and either $p^2|q$ or $p \leqslant 5$. Let $\mathscr{C}$ denote the set of squarefree numbers all of whose prime factors p satisfy $5 < p \leqslant n$. Finally, let $\mathscr{D}$ denote the set of natural numbers with no prime factor exceeding n. Then each q in $\mathscr{D}$ can be written uniquely in the form $q = rs$ with $r \in \mathscr{B}$, $s \in \mathscr{C}$ and $(r, s) = 1$.

The next stage of the argument is to estimate

$$\sum_{\substack{u < q \leqslant V \\ q \in \mathscr{D}}} A(m, q) \tag{8.43}$$

where

$$U = n^{1/12}, \quad V = \exp((\log n)^{1+2\delta}). \tag{8.44}$$

By (8.5) and Lemma 4.5, $A(m, q)$ is a multiplicative function of q. Therefore

$$\left|\sum_{\substack{U < q \leqslant V \\ q \in \mathscr{D}}} A(m, q)\right| \leqslant \sum_{\substack{r > n^{80\delta} \\ r \in \mathscr{B}}} \sum_{s \in \mathscr{C}} |A(m, r)A(m, s)|$$
$$+ \sum_{\substack{r \leqslant n^{80\delta} \\ r \in \mathscr{B}}} |A(m, r)| \left|\sum_{\substack{U/r < s \leqslant V/r \\ (s, r) = 1, s \in \mathscr{C}}} A(m, s)\right|. \tag{8.45}$$

The first double sum is

$$\leqslant n^{-2\delta}\left(\sum_{r \in \mathscr{B}} r^{1/40}|A(m, r)|\right)\left(\sum_{s \in \mathscr{C}} |A(m, s)|\right) \tag{8.46}$$

The first sum here is

$$\left(\prod_{5<p\leqslant n}\left(1+\sum_{h=2}^{\infty} p^{h/40}|A(m,p^h)|\right)\right)\prod_{p\leqslant 5}\left(1+\sum_{h=1}^{\infty} p^{h/40}|A(m,p^h)|\right)$$

and, by (8.31), (8.37) and (8.38), this is

$$\ll \prod_{p,t}(1+C(t+1))$$

where the product is over all pairs p, t with $p^t \| m$. This is

$$\ll n^{\varepsilon}.$$

By (8.38), the second sum in (8.46) does not exceed

$$\prod_{p\leqslant n}(1+8p^{-1}) \ll n^{\varepsilon}.$$

Hence, by (8.31) and (8.45),

$$\sum_{\substack{U<q\leqslant V\\ q\in\mathscr{D}}} A(m,q) \ll n^{-\delta}+F(m) \tag{8.47}$$

where

$$F(m)=\sum_{r\leqslant n^{80\delta}}\left|\sum_{\substack{U/r<s\leqslant V/r\\ (s,r)=1, s\in\mathscr{C}}} A(m,s)\right|. \tag{8.48}$$

By (8.39) and (8.40) and the multiplicative property of $A(m,s)$,

$$A(m,s)=\sum_{\substack{\chi\\ \bmod s}}^{*} c(\chi)\chi(m) \quad (s\in\mathscr{C}) \tag{8.49}$$

where $\sum^*$ denotes a sum over primitive characters,

$$|c(\chi)|\leqslant s^{-1} \tag{8.50}$$

and, for $\lambda>0$,

$$\sum_{\substack{\chi\\ \bmod s}}^{*} |c(\chi)|^{\lambda} \leqslant 8^{\omega(s)} s^{-\lambda}. \tag{8.51}$$

Lemma 8.2 *Let l denote a natural number. Then, for arbitrary complex numbers $b(\chi)$,*

$$\left(\sum_{x=1}^{N}\left|\sum_{q\leqslant Q}\sum_{\substack{\chi\\ \bmod q}}^{*} b(\chi)\chi(x)\right|^{3/2}\right)^{2/3}$$
$$\ll B\left(\sum_{q\leqslant Q}\sum_{\substack{\chi\\ \bmod q}}^{*} |b(\chi)|^{2l/(2l-1)}\right)^{(2l-1)/2l}$$

where

$$B = (N^{1/2} + Q^{1/l})N^{1/6}(\log(N^l e))^{(l^4-1)/(6l)}$$

and the implied constant is absolute.

Proof By the case $l = 1$ of Lemma 5.3 (the large sieve inequality), for arbitrary complex numbers $c_1, \ldots, c_N$,

$$\sum_{q \leqslant Q} \sum_{\substack{a=1 \\ (a,q)=1}}^{q} \left| \sum_{x=1}^{N} c_x e(ax/q) \right|^2 \ll (N + Q^2) \sum_{x=1}^{N} |c_x|^2. \tag{8.52}$$

By the theory of Gauss sums (see § 20 of Hasse (1964) or § 9 of Davenport (1966)) for a primitive character χ modulo q,

$$\tau(\bar{\chi})^{-1} \sum_{y=1}^{q} \bar{\chi}(y) e(yx/q) = \chi(x)$$

where $|\tau(\bar{\chi})|^2 = q$. Therefore

$$\sum_{x=1}^{N} c_x \chi(x) = \tau(\bar{\chi})^{-1} \sum_{y=1}^{q} \bar{\chi}(y) \sum_{x=1}^{N} c_x e(yx/q).$$

Hence

$$\sideset{}{^*}\sum_{\chi \bmod q} \left| \sum_{x=1}^{N} c_x \chi(x) \right|^2 \leqslant q^{-1} \sum_{\chi \bmod q} \left| \sum_{y=1}^{q} \bar{\chi}(y) \sum_{x=1}^{N} c_x e(yx/q) \right|^2.$$

Hence, by the orthogonality of characters, and (8.52),

$$\sum_{q \leqslant Q} \sideset{}{^*}\sum_{\chi \bmod q} \left| \sum_{x=1}^{N} c_x \chi(x) \right|^2 \ll (N + Q^2) \sum_{x=1}^{N} |c_x|^2.$$

Applying this to the lth power of $\sum_{x=1}^{N} c_x \chi(x)$ gives

$$\sum_{q \leqslant Q} \sideset{}{^*}\sum_{\chi \bmod q} \left| \sum_{x=1}^{N} c_x \chi(x) \right|^{2l} \ll (N^l + Q^2) \sum_{y} |d_y|^2$$

where

$$d_y = \sideset{}{'}\sum_{x_1 \ldots x_l = y} c_{x_1} \ldots c_{x_l}$$

and $\sum'$ denotes that the summation is restricted to x_j with $x_j \leqslant N$. Suppose that $\lambda > 2$. Then, by two applications of Hölder's inequality,

$$|d_y|^2 \leqslant d_l(y)^{2-2/\lambda} \left(\sideset{}{'}\sum_{x_1 \ldots x_l = y} |c_{x_1} \ldots c_{x_l}|^\lambda \right)^{2/\lambda}$$

and

$$\sum_y |d_y|^2 \leqslant \left(\sum_{y=1}^{N^l} d_l(y)^{(2\lambda-2)/(\lambda-2)} \right)^{1-2/\lambda} \left(\sum_{x=1}^{N} |c_x|^\lambda \right)^{2l/\lambda}$$

where $d_l(y)$ is the number of solutions of $x_1 \ldots x_l = y$ in $x_1, \ldots, x_l$. Hence, by Theorem 288 of Hardy, Littlewood & Pólya (1951),

$$\left(\sum_{x=1}^{N} \left| \sum_{q \leqslant Q} \sum_{\substack{\chi \\ \operatorname{mod} q}}^{*} b(\chi)\chi(x) \right|^{\lambda/(\lambda-1)} \right)^{(\lambda-1)/\lambda}$$

$$\ll B_\lambda \left(\sum_{q \leqslant Q} \sum_{\substack{\chi \\ \operatorname{mod} q}}^{*} |b(\chi)|^{2l/(2l-1)} \right)^{(2l-1)/(2l)}$$

where

$$B_\lambda = (N^l + Q^2)^{1/(2l)} \left(\sum_{y=1}^{N^l} d_l(y)^{(2\lambda-2)/(\lambda-2)} \right)^{(1-2/\lambda)/(2l)}$$

Let $\lambda = 3$. Then the lemma follows provided that, for $X \geqslant 1$,

$$\sum_{y \leqslant X} d_l(y)^4 \ll X (\log Xe)^{l^4-1}.$$

In fact, it is easily seen by induction on r that $d_r(xy) \leqslant d_r(x) d_r(y)$ and by induction on s that

$$\sum_{y \leqslant X} d_r(y)^s \ll X (\log Xe)^{r^s-1}$$

and

$$\sum_{y \leqslant X} d_r(y)^s y^{-1} \ll (\log Xe)^{r^s}.$$

Let $Q_0 = Ur^{-1}$ and $Q_l = n^{l/2}$, let $b(\chi) = c(\chi)$ when q, the modulus of χ, is in $\mathscr{C}$, $(q, r) = 1$ and $U/r < q \leqslant V/r$, and let $b(\chi) = 0$ otherwise. Then, by (8.48) and (8.49),

$$F(m) = \sum_{r \leqslant n^{80\delta}} \left| \sum_q \sum_{\substack{\chi \\ \operatorname{mod} q}}^{*} b(\chi)\chi(m) \right|. \tag{8.53}$$

By Lemma 8.2, Hölder's inequality, (8.50) and (8.51),

$$\sum_{m=n+1}^{2n} \left| \sum_{Q_{l-1} < q \leqslant Q_l} \sum_{\substack{\chi \\ \operatorname{mod} q}}^{*} b(\chi)\chi(m) \right|$$

$$\ll n(l \log(2ne))^{(l^4-1)/(6l)} Q_{l-1}^{-1/(2l)} \times \prod_{p \leqslant n} (1 + 8p^{-1})^{(2l-1)/(2l)}.$$

This is

$$\ll n^{7/8}(l \log (2ne))^{(l^4-1)/(6l)}(\log 2n)^{(8l-4)/l}$$

or

$$\ll nU^{-1/2}r^{1/2}(\log n)^4$$

according as $l > 1$ or $l = 1$. Hence, summing over l with $Q_{l-1} \leqslant V$ gives, via (8.44) and (8.53),

$$\sum_{m=n+1}^{2n} F(m) \ll n^{47/48}.$$

Hence, for all but $\ll n^{1-\delta}$ values of m with $n < m \leqslant 2n$ one has $F(m) \ll n^{-\delta}$, and so, by (8.47), when m is not exceptional,

$$\sum_{\substack{U < q \leqslant V \\ q \in \mathscr{D}}} A(m, q) \ll n^{-\delta}. \tag{8.54}$$

The proof of Theorem 8.3 is completed by examining

$$\sum_{\substack{q > V \\ q \in \mathscr{D}}} A(m, q).$$

Let $\lambda = 1/(\log n)$. Then

$$\sum_{\substack{q > V \\ q \in \mathscr{D}}} |A(m, q)| \leqslant \sum_{q \in \mathscr{D}} (q/V)^{\lambda} |A(m, q)|$$

$$= V^{-\lambda} \prod_{p \leqslant n} \left(1 + \sum_{h=1}^{\infty} p^{h\lambda} |A(m, p^h)|\right).$$

Hence, by (8.31), (8.38) and (8.44),

$$\sum_{\substack{q > V \\ q \in \mathscr{D}}} |A(m, q)|$$

$$\ll \exp(-(\log n)^{2\delta}) \prod_{5 < p \leqslant n} \left(1 + 240 \sum_{k=0}^{\infty} p^{(30\lambda - 1)k + 30\lambda - 1}\right)$$

$$\ll \exp(-(\log n)^{\delta}).$$

Therefore, by (8.54), (8.44) and (8.6), for all but $\ll n^{1-\delta}$ values of m with $n < m \leqslant 2n$ one has

$$\left(\prod_{p \leqslant n} \left(\sum_{h=0}^{\infty} A(m, p^h)\right)\right) - \mathfrak{S}(m, n^{1/12}) \ll \exp(-(\log n)^{\delta}),$$

as required.

8.7 Completion of the proof of Theorem 8.1

By Theorems 8.2 and 8.3, for all but $\ll n^{1-\delta}$ values of m with $n < m \leqslant 2n$ one has

$$R(m) = I(m)\left(\left(\prod_{p \leqslant n}\left(\sum_{h=0}^{\infty} A(m, p^h)\right)\right) + O(\exp(-(\log n)^{\delta}))\right) + O(n^{1/30-\delta}).$$

Consider $I(m)$, given by (8.3), when $n < m \leqslant 2n$. For y_3, y_5 satisfying $\frac{1}{2}m - \frac{1}{4}n < y_3, y_5 < \frac{1}{2}m - \frac{1}{8}n$ one has $\frac{1}{4}n < y_3, y_5 < n$ and $\frac{1}{4}n < m - y_3 - y_5 < n$. Hence $I(m) \gg n^{1/30}$. It is trivial that $I(m) \ll n^{1/30}$. Thus it suffices to show that

$$\prod_{p \leqslant n}\left(\sum_{h=0}^{\infty} A(m, p^h)\right) \gg (\log n)^{-C}. \tag{8.55}$$

By (8.38), there is a constant C such that

$$\prod_{C < p \leqslant n}\left(\sum_{h=0}^{\infty} A(m, p^h)\right) \geqslant \prod_{C < p \leqslant n}(1 - Cp^{-1}) \gg (\log n)^{-C}.$$

Therefore it is only necessary to show that for each prime p one has

$$\sum_{h=0}^{\infty} A(m, p^h) \geqslant p^{-6}. \tag{8.56}$$

It is easily deduced from (8.5) (cf. Lemma 2.12) that

$$p^{2t}\sum_{h=0}^{t} A(m, p^h) = M(m, p^t) \tag{8.57}$$

where $M(m, p^t)$ is the number of solutions of

$$x^2 + y^3 + z^5 \equiv m \pmod{p^t} \tag{8.58}$$

with $1 \leqslant x, y, z \leqslant p^t$. Let $\gamma(2) = 3$, $\gamma(p) = 1$ $(p > 2)$. When $p \nmid a$ the congruence $x^2 \equiv a \pmod{p^t}$ has a solution for each $t \geqslant \gamma(p)$ whenever it has one for $t = \gamma(p)$. Thus if it can be shown that (8.58) is soluble when $t = \gamma(p)$ with $p \nmid x$, then $p^{2t-2\gamma(p)}$ different solutions can be produced in the general case $t \geqslant \gamma(p)$ by taking any y', z' with $y' \equiv y \pmod{p^{\gamma(p)}}$, $z' \equiv z \pmod{p^{\gamma(p)}}$. Thus

$$M(m, p^t) \geqslant p^{2t-2\gamma(p)}$$

which, by (8.57), yields (8.56).

It is trivial that (8.58) is soluble with $2\nmid x$ when $p=2$ and $t=\gamma(p)=3$. It remains to establish the corresponding result when $p>2$.

The number of cubic or zero residues modulo p is at least $(p-1)/(3,p-1)+1$. Hence the conclusion will follow by the pigeon hole principle if it can be shown that the number N of residues modulo p of the form x^2 or x^2+1 with $1\leqslant x\leqslant p-1$ satisfies

$$N=\frac{1}{4}\left(3p+\left(\frac{-1}{p}\right)_L\right).$$

This is readily done starting from the formula

$$N=p-\frac{1}{2}+\frac{1}{2}\left(\frac{-1}{p}\right)_L-\sum_{x=2}^{p-1}\frac{1}{4}\left(1-\left(\frac{x}{p}\right)_L\right)\left(1-\left(\frac{x-1}{p}\right)_L\right).$$

8.8 Exercises

1 Show that almost every natural number is of the form $p+x^k$.

2 Show that card $\{n : n \neq p+x^k,\ n\leqslant X\} \gg X^{1/k}$.

3 Let $R(n)$ denote the number of solutions of

$$x^2+y^3+z^6=n$$

with $x>0$, $y>0$, $z>0$. Show that

(i) $\sum_{n\leqslant X} R(n)=X\Gamma(\frac{3}{2})\Gamma(\frac{4}{3})\Gamma(\frac{7}{6})+O(X^{5/6})$,

(ii) $\Gamma(\frac{3}{2})\Gamma(\frac{4}{3})\Gamma(\frac{7}{6})=0.73\ldots$,

(iii) $x^2+y^3+z^6\equiv n \pmod q$ is always soluble with $(x,q)=1$.

4 Obtain an asymptotic formula for the number of representations of a number as a sum of two squares, two cubes and two fifth powers.

9

Homogeneous equations and Birch's theorem

9.1 Introduction

Let $F(x_1, \ldots, x_s)$ be a homogeneous form of degree $k \geqslant 2$ with integer coefficients. A natural question is to ask whether the equation

$$F(x_1, \ldots, x_s) = 0 \tag{9.1}$$

has a non-trivial solution, i.e. a solution in integers x_j not all zero. Obviously when k is even the equation may only have the trivial solution. However, when k is odd there is more hope. Lewis (1957) building on earlier work of Brauer (1945) showed that if s is sufficiently large, then any cubic form in s variables with integer coefficients has a non-trivial zero. Shortly afterwards this was extended by Birch (1957) to forms of arbitrary odd degree. Indeed, Birch proved somewhat more than this. The object here is to give an account of Birch's theorem. For references to later work on this and related topics the interested reader should see Davenport's collected works (Davenport, 1977).

The proof of Birch's theorem rests on a special case, namely on the solubility of the additive homogeneous equation

$$c_1 x_1^k + \ldots + c_s x_s^k = 0, \tag{9.2}$$

and this can be treated by an application of the Hardy–Littlewood method.

9.2 Additive homogeneous equations

The methods of Chapters 2, 4 and 5 are readily adapted to give the following theorem, and so the proof is only given in outline.

Theorem 9.1 *Let $k \geqslant 2$ and s_0 be as in Theorem 5.4, and suppose that $s \geqslant \min(s_0, 2^k + 1)$ and $s \geqslant 4k^2 - k + 1$. Suppose further that when k*

is even not all of the integers $c_1, \ldots, c_s$ are of the same sign. Then the equation (9.2) *has a non-trivial solution in integers $x_1, \ldots, x_s$.*

Throughout this section, implicit constants may depend on $c_1, \ldots, c_s$.

If there is a j such that $c_j = 0$, then the conclusion is trivial. Thus it may be assumed that, for every j, $c_j \neq 0$. Also, when k is odd, it can be assumed (if necessary by replacing x_1 by $-x_1$) that not all the c_j are of the same sign. Let $R(N)$ denote the number of solutions of (9.2) with $1 \leqslant x_j \leqslant N$. Then the methods developed in Chapters 2, 4 and 5 give

$$R(N) = \mathfrak{S}J(N) + O(N^{s-k-\delta})$$

where

$$\mathfrak{S} = \prod_p T(p), \quad T(p) = \sum_{h=0}^{\infty} S(p^h),$$

$$S(q) = \sum_{\substack{a=1 \\ (a,q)=1}}^{q} \prod_{j=1}^{s} (q^{-1} S(q, ac_j)),$$

and

$$J(N) = \sum_{\substack{m_1=1 \\ c_1 m_1 + \ldots + c_s m_s = 0}}^{N^k} \ldots \sum_{m_s=1}^{N^k} k^{-s}(m_1 \ldots m_s)^{1/k-1}.$$

They further show that there is a number C, depending at most on $c_1, \ldots, c_s$, such that

$$\prod_{p>C} T(p) > \tfrac{1}{2}$$

and that

$$J(N) \gg N^{s-k}.$$

Now it suffices to show that $T(p) > 0$ and, again, this will follow if it is shown that $M_F(q)$, the number of solutions of

$$F(x_1, \ldots, x_s) = c_1 x_1^k + \ldots + c_s x_s^k \equiv 0 \pmod q$$

with $1 \leqslant x_j \leqslant q$, satisfies, for t sufficiently large,

$$M_F(p^t) > C(p) p^{t(s-1)} \tag{9.3}$$

for some positive number $C(p)$ depending only on $c_1, \ldots, c_s$ and p.

In order to treat M_F it is necessary to transform the variables so as to obtain a new form H in which an appreciable number of the coefficients are coprime with p. Choose τ_j so that $p^{\tau_j} \| c_j$ and choose h_j, l_j so that $\tau_j = h_j k + l_j$ and $0 \leqslant l_j < k$. Then

$$F(x_1, \ldots, x_s) = G(p^{h_1}x_1, \ldots, p^{h_s}x_s)$$

where

$$G(x_1, \ldots, x_s) = d_1 p^{l_1} x_1^k + \ldots + d_s p^{l_s} x_s^k$$

with $d_j = c_j p_j^{-\tau_j}$. Now let $h = \max h_j$. Then

$$F(p^{h-h_1}x_1, \ldots, p^{h-h_s}x_s) = p^{hk}G(x_1, \ldots, x_s)$$

and, for $t > h$,

$$M_F(p^t) \geqslant \sum_{\substack{x_1 = 1}}^{pt - h + h_1} \ldots \sum_{\substack{x_s = 1 \\ p^{hk}G(x_1,\ldots,x_s) \equiv 0 (\bmod p^t)}}^{pt - h + h_s} 1$$

$$\geqslant M_G(p^{t-hk}) \prod_{j=1}^{s} p^{hk - h + h_j},$$

whence

$$M_F(p^t) \geqslant M_G(p^{t-hk}). \tag{9.4}$$

The form G can be rewritten as

$$G = G^{(0)} + pG^{(1)} + \ldots + p^{k-1}G^{(k-1)}$$

where

$$G^{(j)} = G^{(j)}(\mathbf{x}^{(j)}) = \sum_{\substack{i=1 \\ l_i = j}}^{s} d_i x_i^k.$$

Clearly there exist i and r with $r \geqslant s/k$ and $G^{(i)}$ containing at least r variables. Consider the form

$$H(x_1, \ldots, x_s) = \left(\sum_{j<i} p^j G^{(j)}(p\mathbf{x}^{(j)}) + \sum_{j \geqslant i} p^j G^{(j)}(\mathbf{x}^{(j)}) \right) p^{-i}.$$

Now

$$M_G(p^t) \geqslant M_H(p^{t-i}) \tag{9.5}$$

and H has the shape

$$H = H^{(0)} + pH^{(1)} + \ldots + p^{k-1}H^{(k-1)}$$

with $H^{(0)}$ containing at least r variables, where $r \geqslant s/k$, and all its coefficients relatively prime to p. It can be assumed, if necessary by relabelling the variables, that

$$H^{(0)} = H^{(0)}(x_1, \ldots, x_r) = d_1x_1^k + \ldots + d_rx_r^k.$$

By (9.5) and (9.4), to prove (9.3) it now suffices to show that there is a positive number $C_1(p)$ such that for t sufficiently large

$$M_H(p^t) > C_1(p)p^{t(s-1)}. \tag{9.6}$$

Let τ denote the highest power of p dividing k and write $\gamma = \tau + 1$ when $p > 2$ or $\tau = 0$, and $\gamma = \tau + 2$ when $p = 2$ and $\tau \geqslant 1$. Then, as in §2·6, (9.6) will follow on showing that, for each m,

$$d_1x_1^k + \ldots + d_rx_r^k \equiv m \pmod{p^\gamma} \tag{9.7}$$

is soluble in $x_1, \ldots, x_r$ with $p \nmid x_1$.

Let $K = p^{\gamma - \tau - 1}(k, p^\tau(p-1))$. Then the number of kth power residues modulo p^γ is $\phi(p^\gamma)/K$. Hence, by Lemma 2.14, the set $\mathscr{M}_j$ of residues m modulo p^γ which can be written in the form

$$d_1x_1^k + \ldots + d_jx_j^k \quad (p \nmid x_1)$$

satisfies card $\mathscr{M}_j \geqslant \min(p^\gamma, j\phi(p^\gamma)/K)$. Thus, if $r \geqslant 4k$, i.e. $s > 4k^2 - k$, then (9.7) has a solution of the desired kind, and this completes the proof of Theorem 9.1.

Suppose that $c_1, \ldots, c_s$ are integers such that for every q the congruence

$$c_1x_1^k + \ldots + c_sx_s^k \equiv 0 \pmod{q}$$

has a solution with $(x_j, q) = 1$ for some j. Then, following Davenport & Lewis (1963) $c_1, \ldots, c_s$ are said to satisfy *the congruence condition.* They define $\Gamma^*(k)$ to be the least s such that every set of s integers $c_1, \ldots, c_s$ satisfies the congruence condition. They further define $G^*(k)$ to be the least number t such that whenever $s \geqslant t$ the equation

$$c_1x_1^k + \ldots + c_sx_s^k = 0$$

has a non-trivial solution in integers when $c_1, \ldots, c_s$ are not all of the same sign when k is even and satisfy the congruence condition.

The argument above gives $\Gamma^*(k) \leqslant 4k^2 - k + 1$ and $G^*(k) \leqslant \min(s_0, 2^k + 1)$. Davenport and Lewis show (i) that $\Gamma^*(k) \leqslant k^2 + 1$, (ii) that $\Gamma^*(k) = k^2 + 1$ when $k + 1$ is prime, and (iii) that $G^*(k) \leqslant k^2 + 1$ when $k \geqslant 18$ and $k \leqslant 6$. Vaughan (1977*b*) has reduced the gap in (iii) by adapting the methods of Chapters 5, 6, 7 when $11 \leqslant k \leqslant 17$.

For small values of k, $\Gamma^*(k)$ is known. (See Bierstedt (1963), Bovey (1974), Dodson (1967), Norton (1966).) Also, following earlier work of Norton (1966) and Chowla & Shimura (1963), Tietäväinen (1971) has shown that

$$\limsup_{k\to\infty} \frac{\Gamma^*(2k+1)}{k\log k} = \frac{2}{\log 2}.$$

9.3 Birch's theorem

Theorem 9.2 (Birch, 1957) *Let j, l denote natural numbers and let $k_1, \ldots, k_j$ be odd natural numbers. Then there exists a number $\Psi_j(k_1, \ldots, k_j, l)$ with the following property. Let $F_1(\mathbf{x}), \ldots, F_j(\mathbf{x})$ denote forms of degrees $k_1, \ldots, k_j$ respectively in $\mathbf{x} = (x_1, \ldots, x_s)$ with rational coefficients. Then, whenever*

$$s \geqslant \Psi_j(k_1, \ldots, k_j, l)$$

there is an l-dimensional vector space V in $\mathbb{Q}^s$ such that for every $\mathbf{x} \in V$

$$F_1(\mathbf{x}) = \ldots = F_j(\mathbf{x}) = 0.$$

The first step in the proof is to establish the case when $j = 1$, F_1 is additive and $k \geqslant 3$.

Lemma 9.1 *There is a number $\Phi(k, l)$, defined for natural numbers k, l with k odd and $k \geqslant 3$, such that, if $s \geqslant \Phi(k, l)$, then for each form $c_1x_1^k + \ldots + c_sx_s^k$ with $c_1, \ldots, c_s$ rational, there is an l-dimensional vector space V in $\mathbb{Q}^s$ such that for every $\mathbf{x} \in V$*

$$c_1x_1^k + \ldots + c_sx_s^k = 0. \tag{9.8}$$

Proof By Theorem 9.1 there are $t = t(k)$ and $y_1, \ldots, y_t$ not all zero such that

$$c_1y_1^k + \ldots + c_ty_t^k = 0.$$

Similarly for

$$c_{t+1}y_{t+1}^k + \ldots + c_{2t}y_{2t}^k = 0$$

and so on. Hence, when $s \geqslant lt$, the point

$$(u_1y_1, \ldots, u_1y_t, u_2y_{t+1}, \ldots, u_2y_{2t}, \ldots, u_ly_{lt}, 0, \ldots, 0)$$

satisfies (9.8) for all $u_1, \ldots, u_l$.

Proof of Theorem 9.2 Let $k = \max k_i$, so that k is an odd positive integer. The proof is by induction through odd values of k. The result for $k = 1$ is straightforward. For $k \geqslant 3$ the principal step is to show that if the result holds for systems of forms with $\max k_i \leqslant k - 2$, then it holds for a single form of degree k. The conclusion is then easily extended to a system of forms of degree at most k.

For a form

$$F(\mathbf{x}) = F(x_1, \ldots, x_s) = \sum_{i_1, \ldots, i_k} c_{i_1, \ldots, i_k} x_{i_1} \ldots x_{i_k}$$

of (odd) degree k, consider

$$\begin{aligned} &F(u_0 \mathbf{y}^{(0)} + u_1 \mathbf{y}^{(1)} + \ldots + u_{n+1} \mathbf{y}^{(n+1)}) \\ &\quad = \sum_{i_1, \ldots, i_k} c_{i_1, \ldots, i_k} (u_0 y_{i_1}^{(0)} + \ldots + u_{n+1} y_{i_1}^{(n+1)}) \ldots \\ &\qquad \ldots (u_0 y_{i_k}^{(0)} + \ldots + u_{n+1} y_{i_k}^{(n+1)}) \\ &\quad = \sum_{\substack{j_1, \ldots, j_k \\ 0 \leqslant j_r \leqslant n+1}} u_{j_1} \ldots u_{j_k} \sum_{i_1, \ldots, i_k} c_{i_1, \ldots, i_k} y_{i_1}^{(j_1)} \ldots y_{i_k}^{(j_k)}. \end{aligned}$$

Now define $e^{(1)} = (1, 0, 0, \ldots)$, $e^{(2)} = (0, 1, 0, \ldots)$, and so on, and take $u_0 = v$, $\mathbf{y}^{(0)} = \mathbf{y}$, $\mathbf{y}^{(1)} = \mathbf{e}^{(1)}$, $\mathbf{y}^{(2)} = \mathbf{e}^{(2)}, \ldots$. Then a further regrouping of terms gives

$$\begin{aligned} &F(v\mathbf{y} + u_1 \mathbf{e}^{(1)} + \ldots + u_{n+1} \mathbf{e}^{(n+1)}) \\ &\quad = \sum_{h=0}^{k} v^h \sum_{\substack{j_1, \ldots, j_{k-h} \\ 1 \leqslant j_r \leqslant n+1}} u_{j_1} \ldots u_{j_{k-h}} F(\mathbf{y}; h, j_1, \ldots, j_{k-h}) \end{aligned} \tag{9.9}$$

where

$$F(\mathbf{y}; h, j_1, \ldots, j_{k-h})$$

is a form of degree h in $\mathbf{y} = (y_1, \ldots, y_s)$. The total number of such forms with h odd, $1 \leqslant h \leqslant k - 2$ and $1 \leqslant j_r \leqslant n + 1$ does not exceed $k(n+1)^k$. Hence, on the inductive hypothesis, and provided that

$$s \geqslant \Psi_{k(n+1)^k}(k-2, \ldots, k-2, 1),$$

one finds that the corresponding simultaneous equations

$$F(\mathbf{y}; h, j_1, \ldots, j_{k-h}) = 0$$

have a non-trivial solution $\mathbf{z}^{(0)}$ in $\mathbb{Q}^s$.

If $\mathbf{z}^{(0)}, \mathbf{e}^{(1)}, \ldots, \mathbf{e}^{(n+1)}$ are linearly dependent over $\mathbb{Q}$, then omitting one of the $e^{(j)}$ gives a linearly independent set of $n+1$ points in $\mathbb{Q}^s$. Thus, in any case, by taking one of the u_j to be zero in (9.9) and, if necessary, relabelling, one obtains $z^{(0)}, z^{(1)}, \ldots, z^{(n)}$ that are linearly independent and such that

$$F(vz^{(0)} + u_1 z^{(1)} + \ldots + u_n z^{(n)}) = cv^k + \sum_{\substack{h=2 \\ h \text{ even}}}^{k-1} v^h G_h(u) + G_0(u) \qquad (9.10)$$

where $G_h(u)$ is a form of degree $k-h$ in $u = (u_1, \ldots, u_n)$.

The linear independence of $z^{(0)}, \ldots, z^{(n)}$ ensures that, when

$$x = vz^{(0)} + u_1 z^{(1)} + \ldots + u_n z^{(n)},$$

non-trivial choices for $(v, u_1, \ldots, u_n)$ give non-trivial values for x.

Consider the system of forms

$$G_h(u) = 0, \quad h \text{ even}, \quad 2 \leqslant h \leqslant k-1. \qquad (9.11)$$

The degree, $k-h$, is odd in each case. Hence, a further application of the inductive hypothesis shows that when $n \geqslant \Psi_k(k-2, \ldots, k-2, m)$, i.e.

$$s \geqslant s_0(k, m),$$

the system (9.11) is soluble for every member u of an m-dimensional vector space U in $\mathbb{Q}^n$. Let $u^{(1)}, \ldots, u^{(m)}$ denote m linearly independent points in U and consider

$$u = w_1 u^{(1)} + \ldots + w_m u^{(m)}.$$

The linear independence again ensures that non-trivial w in $\mathbb{Q}^m$ give rise to non-trivial u in $\mathbb{Q}^n$. Hence, by (9.10), for non-trivial $(v, w_1, \ldots, w_m)$ there are non-trivial $x = (x_1, \ldots, x_s)$ such that

$$F(x) = cv^k + H(w)$$

where H is a form in $w = (w_1, \ldots, w_m)$ of degree k, i.e. F represents $cv^k + H(w)$.

Continued repetition of this argument shows that if $s \geqslant s_1(k, l)$, then F represents a diagonal form

$$c_1 v_1^k + \ldots + c_t v_t^k$$

with $t = \Phi(k, l)$. Lemma 9.1 now gives the case $j=1$, $k_1 = k$ of the theorem.

To complete the inductive argument, it remains to establish the general case of j simultaneous equations $F_1 = \ldots = F_j = 0$ with

$\max k_i = k$. This is done by subinduction on j. The case $j = 1$ has just been dealt with. Suppose $j > 1$. Without loss of generality it can be supposed that $k_j = k$. By the case $j = 1$, given m, if $s \geqslant \Psi_1(k_j, m)$, then there is an m-dimensional vector space U in $\mathbb{Q}^s$ such that $F_j(\mathbf{x}) = 0$ for every $\mathbf{x}$ in U. The points of U can be represented by

$$y_1\mathbf{x}^{(1)} + \ldots + y_m\mathbf{x}^{(m)}$$

where $\mathbf{x}^{(1)}, \ldots, \mathbf{x}^{(m)}$ are linearly independent points of $\mathbb{Q}^s$. For these points the forms $F_1, \ldots, F_{j-1}$ become forms in $\mathbf{y} = (y_1, \ldots, y_m)$. If

$$\max_{1 \leqslant i \leqslant j-1} k_i \leqslant k - 2$$

then one uses the main inductive hypothesis. If

$$\max_{1 \leqslant i \leqslant j-1} k_i = k$$

then one uses instead the subinductive hypothesis. In either case, provided that $m \geqslant \Psi_{j-1}(k_1, \ldots, k_{j-1}, l)$, there is an l-dimensional vector space V in $\mathbb{Q}^m$ on which each F_i vanishes. This completes the proof of the theorem.

9.4 Exercises

1 Adapt the methods of Chapter 7 to show that

$$\limsup_{k \to \infty} \frac{G^*(k)}{k \log k} \leqslant 2.$$

2 Adapt the methods of Chapter 6 to show that $G^*(3) \leqslant 8$, $G^*(4) \leqslant 14$, $G^*(5) \leqslant 23$, $G^*(6) \leqslant 36$.

3 Show that $\Gamma^*(2) = 5$, $\Gamma^*(3) = 7$, $\Gamma^*(4) = 17$, and that $\Gamma^*(k) \leqslant \min(s_0, 2^k + 1)$.

10

A theorem of Roth

10.1 Introduction

van der Waerden (1927) proved that given natural numbers l, r there exists an $n_0(l, r)$ such that if $n \geqslant n_0(l, r)$ and $\{1, 2, \ldots, n\}$ is partitioned into r sets, then at least one set contains l terms in arithmetic progression.

For an arbitrary set $\mathscr{A}$ of natural numbers, let

$$A(n) = A(n, \mathscr{A}) = \sum_{\substack{a \leqslant n \\ a \in \mathscr{A}}} 1, \quad D(n) = D(n, \mathscr{A}) = \frac{1}{n} A(n) \tag{10.1}$$

and write $\underline{d}$ and $\bar{d}$ for the lower and upper asymptotic densities of $\mathscr{A}$,

$$\underline{d} = \underline{d}(\mathscr{A}) = \liminf_{n \to \infty} D(n) \quad \text{and} \quad \bar{d} = \bar{d}(\mathscr{A}) = \limsup_{n \to \infty} D(n) \tag{10.2}$$

respectively. When $\underline{d} = \bar{d}$ let $d = d(\mathscr{A})$ denote their common value, the asymptotic density of $\mathscr{A}$. Erdős & Turán (1936), in discussing the nature of the known proofs of van der Waerden's theorem, conjectured that every set $\mathscr{A}$ with $\bar{d}(\mathscr{A}) > 0$ contains arbitrarily long arithmetic progressions. An equivalent assertion is that if there is an l such that $\mathscr{A}$ contains no arithmetic progression of l terms, then $d(\mathscr{A}) = 0$.

The first non-trivial case is $l = 3$. The initial breakthrough was made by Roth (1952, 1953, 1954) in establishing this case by an ingenious adaptation of the Hardy–Littlewood method.

By a different method, Szemerédi (1969) proved the conjecture for $l = 4$, and Roth (1972) has given an alternative proof by an approach related to that of his earlier method.

In 1975, Szemerédi established the general case. Unfortunately Szemerédi's proof uses van der Waerden's theorem. More recently Furstenburg (1977) has given a proof of Szemerédi's theorem based on ideas from ergodic theory. Although this does not use van der Waerden's theorem it apparently has a similar structure and so still does not yield the sought after insight.

Ideas stemming from the attacks on this problem have enabled Furstenburg (1977) and Sárközy (1978*a*, *b*) to establish that if $\bar{d}(\mathscr{A}) > 0$, then the set of numbers of the form $a - a'$ with $a \in \mathscr{A}, a' \in \mathscr{A}$ contains infinitely many perfect squares.

In this chapter, Roth's theorem is established using his version of the Hardy–Littlewood method, and a proof of the Sárközy–Furstenburg theorem is developed along the lines of Furstenburg but avoiding the ergodic theory.

Throughout this chapter implicit constants are absolute.

10.2 Roth's theorem

Let $M^{(l)}(n)$ denote the largest number of elements which can be taken from $\{1, 2, \ldots, n\}$ with no l of them in progression. Let

$$\mu^{(l)}(n) = n^{-1} M^{(l)}(n).$$

Then Szemerédi's theorem is the assertion $\lim_{n \to \infty} \mu^{(l)}(n) = 0$, and this obviously implies the Erdős–Turán conjecture. As the following lemma shows, it is quite easy to prove that the limit exists. Its value is another matter.

Lemma 10.1 *For each integer* l, $\lim_{n \to \infty} \mu^{(l)}(n)$ *exists. Also, for* $m \geqslant n$ *one has* $\mu^{(l)}(m) \leqslant 2\mu^{(l)}(n)$.

Proof It is a trivial consequence of the definition of $M^{(l)}$ that

$$M^{(l)}(m+n) \leqslant M^{(l)}(m) + M^{(l)}(n).$$

Hence

$$\begin{aligned} M^{(l)}(m) &\leqslant \left[\frac{m}{n}\right] M^{(l)}(n) + M^{(l)}\left(m - n\left[\frac{m}{n}\right]\right) \\ &\leqslant \frac{m}{n} M^{(l)}(n) + n \end{aligned}$$

Therefore $\mu^{(l)}(m) \leqslant \mu^{(l)}(n) + n/m$, so that

$$\limsup_{m \to \infty} \mu^{(l)}(m) \leqslant \mu^{(l)}(n)$$

whence

$$\limsup_{m \to \infty} \mu^{(l)}(m) \leqslant \liminf_{n \to \infty} \mu^{(l)}(n).$$

Also, when $m \geqslant n$, $M^{(l)}(m) \leqslant (m/n + 1) M^{(l)}(n) \leqslant 2M^{(l)}(n) m/n$.

The following theorem not only shows that when $l = 3$ the limit is 0, but gives a bound for the size of $M^{(3)}(n)$.

Theorem 10.1 (Roth) *Let* $n \geqslant 3$. *Then* $\mu^{(3)}(n) \ll (\log\log n)^{-1}$.

It is henceforward supposed that $l = 3$, and for convenience the superscript (l) is dropped.

Choose $\mathcal{M} \subset \{1, 2, \ldots, n\}$ so that card $\mathcal{M} = M(n)$ and no three elements of $\mathcal{M}$ are in progression. Let

$$f(\alpha) = \sum_{m \in \mathcal{M}} e(\alpha m).$$

Then

$$M(n) = \int_0^1 f(\alpha)^2 f(-2\alpha) \mathrm{d}\alpha \tag{10.3}$$

since the right-hand side is the number of solutions of $m_1 + m_2 = 2m_3$ with $m_j \in \mathcal{M}$ and, by the construction of $\mathcal{M}$, such solutions can only occur when $m_1 = m_2 = m_3$.

Let κ denote the characteristic function of $\mathcal{M}$, so that

$$f(\alpha) = \sum_x \kappa(x) e(\alpha x). \tag{10.4}$$

Suppose that

$$m < n, \tag{10.5}$$

and consider

$$v(\alpha) = \mu(m) \sum_{x=1}^{n} e(\alpha x) \tag{10.6}$$

and

$$E(\alpha) = v(\alpha) - f(\alpha).$$

Then

$$E(\alpha) = \sum_{x=1}^{n} c(x) e(\alpha x) \tag{10.7}$$

with

$$c(x) = \mu(m) - \kappa(x). \tag{10.8}$$

The idea of the proof is that, if $M(n)$ is close to n, then

$$\int_0^1 f(\alpha)^2 f(-2\alpha) \mathrm{d}\alpha$$

ought to be closer to $M(n)^2$ than to $M(n)$ (cf. (10.3)). To show this, one first of all uses the disorderly arithmetical structure of $\mathscr{M}$ to replace f by v with a relatively small error. It is a fairly general principle, observable from the applications of the method in previous chapters, that sums of the form

$$\sum_{\substack{x \leqslant n \\ x \in \mathscr{A}}} e(\alpha x)$$

tend to have large peaks at a/q when the elements of $\mathscr{A}$ are regularly distributed in residue classes modulo q. Note that $v(\alpha)$ has its peaks at the integers.

Let

$$F(\alpha) = \sum_{z=0}^{m-1} e(\alpha z). \tag{10.9}$$

Lemma 10.2 *Let q be a natural number with $q < n/m$, and for $y = 1, 2, \ldots, n - mq$ let*

$$\sigma(y) = \sigma(y; m, q) = \sum_{x=0}^{m-1} c(y + xq). \tag{10.10}$$

Then

$$\sigma(y) \geqslant 0 \quad (y = 1, 2, \ldots, n - mq) \tag{10.11}$$

and

$$F(\alpha q)E(\alpha) = \sum_{y=1}^{n-mq} \sigma(y) e(\alpha(y + mq - q)) + R(\alpha) \tag{10.12}$$

where $R(\alpha)$ satisfies

$$|R(\alpha)| < 2m^2 q. \tag{10.13}$$

Proof By collecting together the terms in the product FE for which $x + zq = h + mq - q$ one obtains

$$F(\alpha q)E(\alpha) = \sum_{h=1+q-mq}^{n} e(\alpha(h + mq - q)) \times \sum_{\substack{z=0 \\ 1 \leqslant h + q(m-1-z) \leqslant n}}^{m-1} c(h + q(m - 1 - z)).$$

The innermost sum is at most m in absolute value, and so the total contribution from the terms with $h \leqslant 0$ and $h > n - mq$ does not exceed, in modulus, $m(mq + (m-1)q) < 2m^2q$. For the remaining values of h one has $1 \leqslant h + q(m-1-z) \leqslant n$ for all z in the interval $[0, m-1]$. This gives (10.12) and (10.13).

By (10.8) and (10.10),

$$\sigma(y) = M(m) - \sum_{x=0}^{m-1} \kappa(y + xq).$$

Let

$$r = \sum_{x=0}^{m-1} \kappa(y + xq).$$

Then r is the number of elements of $\mathscr{M}$ among $y, y+q, \ldots, y+(m-1)q$. Let these elements be $y + x_1 q, \ldots, y + x_r q$. Then no three are in progression. Hence no three of $x_1, \ldots, x_r$ are in progression. Likewise for $1 + x_1, \ldots, 1 + x_r$. Moreover $1 + x_j \leqslant m$. Hence $r \leqslant M(m)$, which gives (10.11).

Lemma 10.3 *Suppose that* $2m^2 < n$. *Then, for every real number* α,

$$|E(\alpha)| < 2n(\mu(m) - \mu(n)) + 16m^2.$$

Proof By Lemma 2.1, there exist a, q such that $(a, q) = 1$, $1 \leqslant q \leqslant 2m$ and $|\alpha - a/q| \leqslant 1/(2qm)$. Then

$$F(\alpha q) = F(\alpha q - a) = F(\beta)$$

where $|\beta| \leqslant 1/(2m)$. Hence, by (10.9),

$$|F(\alpha q)| = \left|\frac{\sin \pi m \beta}{\sin \pi \beta}\right| \geqslant \frac{2m}{\pi}.$$

Thus, by Lemma 10.2,

$$\begin{aligned} \tfrac{1}{2}m|E(\alpha)| &\leqslant \frac{2}{\pi} m|E(\alpha)| \\ &\leqslant |F(\alpha q)E(\alpha)| \\ &< \sum_{y=1}^{n-mq} \sigma(y) + 2m^2 q \\ &< mE(0) + 8m^3. \end{aligned}$$

Moreover, by (10.7) and (10.8),

$$E(0) = \sum_{x=1}^{n} (\mu(m) - \kappa(x)) = n(\mu(m) - \mu(n)).$$

The lemma follows at once.

Proof of Theorem 10.1. Let

$$I = \int_0^1 f(\alpha)^2 v(-2\alpha)\mathrm{d}\alpha. \tag{10.14}$$

Then, by (10.4) and (10.6),

$$I = \sum_{\substack{a \in \mathscr{M} \\ 2|a+b}} \sum_{b \in \mathscr{M}} \mu(m).$$

Thus, if M_1 is the number of odd elements of $\mathscr{M}$ and M_2 the number of even elements, so that $M_1 + M_2 = M(n)$, then

$$I = \mu(m)(M_1^2 + M_2^2) \geqslant \tfrac{1}{2}\mu(m)M(n)^2. \tag{10.15}$$

By (10.3) and (10.14),

$$|M(n) - I| \leqslant \left(\max_\alpha |E(\alpha)|\right) \int_0^1 |f(\alpha)|^2 \mathrm{d}\alpha.$$

Therefore, by Lemma 10.3 and Parseval's identity, when $2m^2 < n$ one has

$$|M(n) - I| \leqslant (2n(\mu(m) - \mu(n)) + 16m^2)M(n).$$

Hence, by (10.15),

$$\mu(m)\mu(n) \leqslant 4(\mu(m) - \mu(n)) + 34m^2 n^{-1} \quad (2m^2 < n). \tag{10.16}$$

Letting $n \to \infty$ and then $m \to \infty$ shows that $\tau = \lim_{n\to\infty} \mu(n)$ satisfies $\tau^2 \leqslant 0$. To establish the quantitative version of this, let

$$\lambda(x) = \mu(2^{3^x}).$$

By Lemma 10.1, it suffices to show that $\lambda(2\mathrm{x}) \ll x^{-1}$.

By (10.16),

$$\lambda(y)\lambda(y+1) \leqslant 4(\lambda(y) - \lambda(y+1)) + 34 \times 2^{-3^y}.$$

Dividing by $\lambda(y)\lambda(y+1)$, summing over $y = x, x+1, \ldots, 2x-1$ and appealing to Lemma 10.1 gives one

$$x \leqslant 4\lambda(2x)^{-1} + 200x\lambda(2x)^{-2}2^{-3^x}.$$

When $\lambda(2x) > 1/x$ the second term on the right is $< \tfrac{1}{2}x$ for x sufficiently large, so that $\lambda(2x) < 8/x$, which gives the desired conclusion.

10.3 A theorem of Furstenburg and Sárközy

Theorem 10.2 *Let $\mathscr{A}$ be a set of natural numbers with $\overline{d}(\mathscr{A}) > 0$, and let $R(n)$ denote the number of solutions of*

$$a - a' = x^2$$

in a, a', x with $a \in \mathscr{A}$, $a' \in \mathscr{A}$, $a \leqslant n$. Then

$$\limsup_{n \to \infty} R(n) n^{-3/2} > 0.$$

This theorem is somewhat stronger than Theorem 1.2 of Furstenburg (1977). The approach of Sárközy (1978) is different. He adapts the methods of § 10.2 to show that if $a - a' = x^2$ has only trivial solutions, then

$$A(n) \ll n(\log\log n)^{2/3}(\log n)^{-1/3}.$$

Let $\mathscr{N}_0$ denote an infinite set of natural numbers such that

$$\lim_{\substack{n \to \infty \\ n \in \mathscr{N}_0}} n^{-1} A(n) = \overline{d}(\mathscr{A}),$$

let

$$\mathfrak{M}_n(q, a) = \{\alpha : |\alpha - a/q| \leqslant q^{-1} n^{-1/2}\}, \tag{10.17}$$

and let

$$f(\alpha) = \sum_{\substack{a \leqslant n \\ a \in \mathscr{A}}} e(\alpha a).$$

It is necessary to show that f has fairly orderly behaviour on $\mathfrak{M}_n(q, a)$. For $n \geqslant 4$,

$$\int_{\mathfrak{M}_n(q,a)} |f(\alpha)|^2 \mathrm{d}\alpha \leqslant \int_0^1 |f(\alpha)|^2 \mathrm{d}\alpha \leqslant n.$$

Hence

$$\int_{\mathfrak{M}_n(q,a)} |f(\alpha)|^2 n^{-1} \mathrm{d}\alpha$$

is bounded uniformly in q, a, n. Therefore one may choose infinite sets $\mathscr{N}(q, a)$ of natural numbers such that

$$\mathscr{N}(1, 1) = \mathscr{N}(1, 0) \subset \mathscr{N}_0, \quad \mathscr{N}(q+1, 1) \subset \mathscr{N}(q, q-1),$$

$\mathcal{N}(q, a') \subset \mathcal{N}(q, a)$ when $1 \leqslant a < a' \leqslant q - 1$ and $(a', q) = (a, q) = 1$ and

$$\rho(q, a) = \lim_{\substack{n \to \infty \\ n \in \mathcal{N}(q, a)}} \int_{\mathfrak{M}_n(q, a)} |f(\alpha)|^2 n^{-1} \mathrm{d}\alpha$$

exists. Thus, given Q, for all sufficiently large n with $n \in \mathcal{N}(Q, Q-1)$ one has

$$\sum_{q \leqslant Q} \sum_{\substack{a=1 \\ (a,q)=1}}^{q} \rho(q, a) < 1 + \sum_{q \leqslant Q} \sum_{\substack{a=1 \\ (a,q)=1}}^{q} \int_{\mathfrak{M}_n(q, a)} |f(\alpha)|^2 n^{-1} \mathrm{d}\alpha$$

and the $\mathfrak{M}_n(q, a)$ with $1 \leqslant a \leqslant q \leqslant Q$ and $(a, q) = 1$ pairwise disjoint. Hence

$$\sum_{q \leqslant Q} \sum_{\substack{a=1 \\ (a,q)=1}}^{q} \rho(q, a) < 1 + \int_0^1 |f(\alpha)|^2 n^{-1} \mathrm{d}\alpha \leqslant 2.$$

Therefore

$$\sum_{q=1}^{\infty} \sum_{\substack{a=1 \\ (a,q)=1}}^{q} \rho(q, a)$$

converges.

10.4 The definition of major and minor arcs

Suppose that $0 < \eta < 1$ and choose $Q = Q_0(\eta)$ so that

$$\sum_{q=Q+1}^{\infty} \sum_{\substack{a=1 \\ (a,q)=1}}^{q} \rho(q, a) < \eta \quad \text{and} \quad Q > \frac{1}{\eta}. \tag{10.18}$$

Now define

$$k = (Q!)^2, \quad P = k^{100} \tag{10.19}$$

and henceforth suppose that

$$n \in \mathcal{N}(P, P-1).$$

Then, given

$$X \geqslant 1$$

choose

$$n_0 = n_0(\eta, X) \geqslant X^2$$

so that when

$$n \geqslant n_0$$

and $1 \leqslant a \leqslant q \leqslant P$ with $(a, q) = 1$ the major arcs

$$\mathfrak{M}_{n,X}(q, a) = \{\alpha : |\alpha - a/q| \leqslant Xq^{-1}n^{-1}\}$$

are pairwise disjoint,

$$\int_{\mathfrak{M}_n(q,a)} |f(\alpha)|^2 n^{-1} \mathrm{d}\alpha < \rho(q, a) + \eta P^{-2} \tag{10.20}$$

and

$$A(n) > \tfrac{2}{3}\bar{d}n. \tag{10.21}$$

By (10.17), $\mathfrak{M}_{n,X}(q, a) \subset \mathfrak{M}_n(q, a)$ for $n \geqslant n_0$. Hence, by (10.20),

$$\int_{\mathfrak{M}_{n,X}(q,a)} |f(\alpha)|^2 n^{-1} \mathrm{d}\alpha < \rho(q, a) + \eta P^{-2}. \tag{10.22}$$

Let $\mathfrak{M}$ denote the union of the major arcs $\mathfrak{M}_{n,X}(q,a)$ with $1 \leqslant a \leqslant q \leqslant P$ and $(a, q) = 1$, and define the minor arcs $\mathfrak{m}$ by

$$\mathfrak{m} = (Xn^{-1}, 1 + \mathrm{X}\mathrm{n}^{-1}] \setminus \mathfrak{M}.$$

Further define

$$g(\beta) = \sum_{x=1}^{N} e(\beta x^2) \tag{10.23}$$

where

$$N \leqslant (n/k)^{1/2}. \tag{10.24}$$

Then, by (10.19),

$$R(n) \geqslant \mathscr{R} \quad \text{where} \quad \mathscr{R} = \int_0^1 g(k\alpha)|f(\alpha)|^2 \mathrm{d}\alpha. \tag{10.25}$$

By Theorem 4.1, when $(a, q) = 1$,

$$g(\gamma) = q^{-1}S(q, a)h\left(\gamma - \frac{a}{q}\right) + O\left(q^{9/16}\left(1 + N^2\left|\gamma - \frac{a}{q}\right|\right)\right) \tag{10.26}$$

where

$$S(q, a) = \sum_{x=1}^{q} e(ax^2/q), \quad h(\beta) = \int_0^{N^2} \tfrac{1}{2}\alpha^{-1/2} e(\beta\alpha)\mathrm{d}\alpha. \tag{10.27}$$

Also, by Theorem 4.2, $S(q, a) \ll q^{1/2}$.

10.5 The contribution from the minor arcs

Suppose $\alpha \in \mathfrak{m}$. Choose a, q so that $(a, q) = 1$, $q \leqslant N^{4/3}$, $|k\alpha - a/q| \leqslant q^{-1}N^{-4/3}$. Let $a_1 = a/(k, a)$, $q_1 = qk/(k, a)$. Then $|\alpha - a_1/q_1| \leqslant q^{-1}N^{-4/3}$ and $(a_1, q_1) = 1$. Since $\alpha \in \mathfrak{m}$, either $q_1 > P$ or $|\alpha - a_1/q_1| > Xn^{-1}q_1^{-1}$. In the first case, by (10.26),

$$g(k\alpha) \ll q^{-1/2}N + q^{9/16}\left(1 + N^2\left|k\alpha - \frac{a}{q}\right|\right)$$
$$\ll Nk^{1/2}P^{-1/2} + N^{3/4},$$

and in the second $|k\alpha - \alpha/q| > Xn^{-1}q^{-1}$, so that, by (10.26), (10.27) and Lemma 2.8,

$$g(k\alpha) \ll q^{-1/2}\left|k\alpha - \frac{a}{q}\right|^{-1/2} + N^{3/4}$$
$$\ll n^{1/2}X^{-1/2} + N^{3/4}.$$

Hence, by (10.19), in either case

$$g(k\alpha) \ll Nk^{-40} + n^{1/2}X^{-1/2} + N^{3/4}.$$

Therefore, by Parseval's identity,

$$\int_{\mathfrak{m}} g(k\alpha)|f(\alpha)|^2 \mathrm{d}\alpha \ll (Nk^{-40} + n^{1/2}X^{-1/2} + N^{3/4})n. \quad (10.28)$$

10.6 The contribution from the major arcs

Now suppose that $\alpha \in \mathfrak{M}_{n,X}(q, a)$ where $1 \leqslant a \leqslant q \leqslant P$ and $(a, q) = 1$. Let $q_1 = q/(q, k)$, $a_1 = ak/(q, k)$. Then, by (10.26),

$$g(k\alpha) = q_1^{-1}S(q_1, a_1)h\left(k\left(\alpha - \frac{a}{q}\right)\right)$$
$$+ O\left(q_1^{9/16}\left(1 + N^2k\left|\alpha - \frac{a}{q}\right|\right)\right).$$

The error term here is majorized by $P + N^2kXn^{-1}$. Hence

$$\int_{\mathfrak{M}} g(k\alpha)|f(\alpha)|^2 \mathrm{d}\alpha = \mathscr{R}_1 + O(Pn + N^2kX) \quad (10.29)$$

where

$$\mathscr{R}_1 = \sum_{q \leqslant P} \sum_{\substack{a=1 \\ (a,q)=1}}^{q} \int_{\mathfrak{M}_{n,X}(q,a)} q_1^{-1}S(q_1, a_1)h\left(k\left(\alpha - \frac{a}{q}\right)\right)|f(\alpha)|^2 \mathrm{d}\alpha.$$

By (10.22) and (10.18) the terms here with $q \geqslant Q+1$ contribute an amount which in absolute value does not exceed

$$\sum_{q=Q+1}^{P} \sum_{\substack{a=1 \\ (a,q)=1}}^{q} Nn(\rho(q,a)+\eta P^{-2}) < 2\eta Nn.$$

Also, when $q \leqslant Q$, by (10.19) one has $q|k$, so that $q_1 = 1$. Hence

$$\mathcal{R}_1 = \mathcal{R}_2 + O(\eta Nn) \tag{10.30}$$

where

$$\mathcal{R}_2 = \sum_{q \leqslant Q} \sum_{\substack{a=1 \\ (a,q)=1}}^{q} \int_{\mathfrak{M}_{n,X}(q,a)} h\left(k\left(\alpha - \frac{a}{q}\right)\right) |f(\alpha)|^2 \mathrm{d}\alpha.$$

It is easily shown that for every positive number Y

$$\int_0^Y \alpha^{-1/2} \cos \alpha \, \mathrm{d}\alpha > 0.$$

Hence, by (10.27),

$$\mathrm{Re}\, h(\beta) = |\beta|^{-1/2} \int_0^{N^2|\beta|} \tfrac{1}{2} \alpha^{-1/2} \cos 2\pi\alpha \, \mathrm{d}\alpha > 0. \tag{10.31}$$

Therefore, on discarding all the terms in $\mathcal{R}_2$ with the exception of that with $a = q = 1$, one obtains

$$\mathrm{Re}\, \mathcal{R}_2 \geqslant \int_{-1/4\pi n}^{1/4\pi n} \mathrm{Re}\, h(k\alpha) |f(\alpha)|^2 \mathrm{d}\alpha.$$

Also, when $|\alpha| \leqslant 1/(4\pi n)$, one has

$$f(\alpha) - f(0) = \sum_{\substack{x=1 \\ x \in \mathscr{A}}}^{n} 2\pi i x \int_0^\alpha e(\beta x) \mathrm{d}\beta$$

so that

$$|f(\alpha)| \geqslant f(0)(1 - 2\pi|\alpha| n) \geqslant \tfrac{1}{2} f(0) = \tfrac{1}{2} A(n).$$

Therefore

$$\mathrm{Re}\, \mathcal{R}_2 \geqslant \tfrac{1}{4} A(n)^2 \int_0^{1/4\pi n} \mathrm{Re}\, h(k\alpha) \mathrm{d}\alpha. \tag{10.32}$$

10.7 Completion of the proof of Theorem 10.2

By (10.24) and (10.31), $\mathrm{Re}\, h(k\alpha) \geqslant \frac{1}{2} N$ whenever $4\pi n|\alpha| \leqslant 1$. Hence, by (10.32) and (10.21),

$$\mathrm{Re}\, \mathcal{R}_2 \geqslant \frac{A(n)^2 N}{32\pi n} > \frac{\bar{d}^2 n N}{250}.$$

Thus, by (10.25), (10.28) and (10.30),

$$R(n) \geqslant \mathscr{R} = \operatorname{Re}\mathscr{R}$$
$$= \operatorname{Re}\mathscr{R}_2 + O((Nk^{-40} + n^{1/2}X^{-1/2} + N^{3/4})n + \eta Nn)$$
$$> \frac{\bar{d}^2}{250}nN - C((Nk^{-40} + n^{1/2}X^{-1/2} + N^{3/4})n + \eta Nn) \tag{10.33}$$

for a suitable constant $C \geqslant 1$.

The proof is completed by making suitable choices of the parameters. Let

$$\eta = 10^{-4}\bar{d}^2C^{-1}.$$

This fixes $Q = Q_0(\eta)$ and so k and P. Note that, by (10.18) and (10.19), $k \geqslant Q > 1/\eta$.

Let

$$X = \eta^{-2}k$$

and suppose that $n \geqslant n_0(\eta, X)$ with $n \in \mathscr{N}(P, P-1)$. Finally, let $N = [(n/k)^{1/2}]$, so that (10.24) holds. Now for $n \geqslant n_1(\eta)$,

$$C((Nk^{-40} + n^{1/2}X^{-1/2} + N^{3/4})n + \eta Nn)$$
$$< C(\eta Nn + \eta n^{3/2}k^{-1/2} + \eta Nn + \eta Nn)$$
$$< 5C\eta Nn$$
$$= \frac{1}{2000}\bar{d}^2 Nn.$$

Hence, by (10.33),

$$\limsup_{n\to\infty} R(n)n^{-3/2} \geqslant \frac{1}{300}\bar{d}^2k^{-1/2} > 0$$

as required.

10.8 Exercises

1 Prove the theorem of Sárközy stated in § 10.3.

2 Show that if $\bar{d}(\mathscr{A}) > 0$ and $R(n)$ denotes the number of solutions of $a - a' = p - 1$ with $a \in \mathscr{A}$, $a' \in \mathscr{A}$, $a \leqslant n$, p prime, then

$$\limsup_{n\to\infty} R(n)(\log n)n^{-2} > 0.$$

11

Diophantine inequalities

11.1 A theorem of Davenport and Heilbronn

All of the forms of the Hardy–Littlewood method described so far have dealt with the solution of equations in integers. For instance, in Chapter 9 it was shown that if s is large enough, then given integers $c_1, \ldots, c_s$ (or equivalently given that $c_1, \ldots, c_s$ are all in rational ratio), not all of the same sign when k is even, the equation

$$c_1x_1^k + \ldots + c_sx_s^k = 0$$

has a non-trivial solution in integers $x_1, \ldots, x_s$. Now one can ask what happens when the $c_1, \ldots, c_s$ are not in rational ratio. It is no longer sensible to insist that the form represents 0, but one can ask instead that it take arbitrarily small values.

In order to answer this question, Davenport & Heilbronn (1946) introduced an important variant of the Hardy–Littlewood method. This enabled them to establish the following theorem.

Theorem 11.1 *Suppose that* $s \geqslant 2^k + 1$ *and that* $\lambda_1, \ldots, \lambda_s$ *are non-zero real numbers not all in rational ratio, and not all of the same sign when* k *is even. Then for every positive number* η *there exist integers* $x_1, \ldots, x_s$, *not all zero, such that*

$$|\lambda_1x_1^k + \ldots + \lambda_sx_s^k| < \eta. \tag{11.1}$$

It suffices to prove the theorem when $\eta = 1$, for it can then be applied with λ_j replaced by λ_j/η. Moreover, when k is odd, replacing, if necessary, x_1^k by $(-x_1)^k$ enables one to assume in this case also that not all the λ_j are of the same sign.

By relabelling it can be supposed that λ_1/λ_2 is irrational. If $\lambda_1/\lambda_2 > 0$, then consider any j for which $\lambda_1/\lambda_j < 0$. Then, when λ_1/λ_j is rational, λ_2/λ_j is irrational and negative. In any case, by further relabelling it can be supposed that

$$\lambda_1/\lambda_2 \text{ is irrational and negative.} \tag{11.2}$$

In all of the forms of the Hardy–Littlewood method considered so far, the line of attack has been via fourier transforms on the torus $\mathbb{T} = \mathbb{R}/\mathbb{Z}$. For the present problem it is more appropriate to work on $\mathbb{R}$. The obvious analogue of (1.8) is the identity.

$$\int_{-\infty}^{\infty} e(\alpha\beta)\frac{\sin 2\pi\alpha}{\pi\alpha}\mathrm{d}\alpha = \begin{cases} 1 & (|\beta| < 1), \\ 0 & (|\beta| > 1). \end{cases}$$

However there are difficulties associated with this transform because the integral does not converge absolutely. It is more convenient, therefore, to use instead

$$I(\beta) = \int_{-\infty}^{\infty} e(\alpha\beta)K(\alpha)\mathrm{d}\alpha, \quad K(\alpha) = \left(\frac{\sin \pi\alpha}{\pi\alpha}\right)^2. \tag{11.3}$$

A straightforward application of the Cauchy integral formula gives

$$I(\beta) = \max(1 - |\beta|, 0). \tag{11.4}$$

Let

$$f(\alpha) = \sum_{x=1}^{N} e(\alpha x^k), \quad f_j(\alpha) = f(\lambda_j\alpha). \tag{11.5}$$

Then for the method to be successful one requires a positive lower bound for

$$R(N) = \int_{-\infty}^{\infty} \left(\prod_{j=1}^{s} f_j(\alpha)\right)K(\alpha)\mathrm{d}\alpha, \tag{11.6}$$

for by (11.4) and (11.5) this is

$$\sum_{x_1=1}^{N} \dots \sum_{x_s=1}^{N} \max(1 - |\lambda_1 x_1^k + \dots + \lambda_s x_s^k|, 0)$$

which can only be positive if there are $x_1, \dots, x_s$ for which (11.1) holds with $\eta = 1$. Thus Theorem 11.1 follows from

Theorem 11.2 *Suppose that $s > 2^k$. Then there are arbitrarily large N for which*

$$R(N) \gg N^{s-k}.$$

Note that throughout this chapter implicit constants may depend on $\lambda_1, \dots, \lambda_s$.

11.2 The definition of major and minor arcs

The form of the Hardy–Littlewood method used here is somewhat simpler than that described hitherto. The most important simplification arises from the fact that for suitable choices of N the integrand has only one really big peak, that at the origin. This is because the irrationality of λ_1/λ_2 ensures that one of f_1, f_2 is relatively small when α is not near the origin.

Let

$$\nu = \frac{1}{100}, \quad P = N^\nu. \tag{11.7}$$

Then $\mathbb{R}$ is divided into three regions. These consist of the sole major arc

$$\mathfrak{M} = \{\alpha : |\alpha| \leqslant PN^{-k}\}, \tag{11.8}$$

the pair of minor arcs

$$\mathfrak{m} = \{\alpha : PN^{-k} < |\alpha| \leqslant P\}. \tag{11.9}$$

and the 'trivial' region

$$\mathfrak{t} = \{\alpha : |\alpha| > P\}. \tag{11.10}$$

The trivial region can be dismissed quickly. By Hua's lemma (Lemma 2.5),

$$\int_X^{X+1} |f_j(\alpha)|^{2^k} \mathrm{d}\alpha \ll N^{2^k - k + \varepsilon},$$

so that, by Hölder's inequality,

$$\int_X^{X+1} \left| \prod_{j=1}^{2^k} f_j(\alpha) \right| \mathrm{d}\alpha \ll N^{2^k - k + \varepsilon}. \tag{11.11}$$

Thus, by (11.3),

$$\int_{\mathfrak{t}} \left| \prod_{j=1}^{s} f_j(\alpha) \right| K(\alpha)\mathrm{d}\alpha \leqslant \int_P^\infty \left| \prod_{j=1}^{s} f_j(\alpha) \right| \alpha^{-2} \mathrm{d}\alpha$$

$$\ll N^{s-k+\varepsilon} \sum_{h=0}^{\infty} (h+P)^{-2}.$$

Therefore

$$\int_{\mathfrak{t}} \left| \prod_{j=1}^{s} f_j(\alpha) \right| K(\alpha)\mathrm{d}\alpha \ll N^{s-k-\delta} \tag{11.12}$$

where here, and below, δ is a fixed positive number depending at most on $k, s, \lambda_1, \ldots, \lambda_s$.

11.3 The treatment of the minor arcs

It is on $\mathfrak{m}$ that use is made of the irrationality of λ_1/λ_2, and the argument requires a specialization of N.

Lemma 11.1 *Let a, q be any pair with $(a, q) = 1$ and*

$$\left|\frac{\lambda_1}{\lambda_2} - \frac{a}{q}\right| \leqslant q^{-2}.$$

Further let $N = q^2$. Then

$$\sup_{\alpha \in \mathfrak{m}} \min(|f_1(\alpha)|, |f_2(\alpha)|) \ll N^{1-\delta}.$$

The existence of arbitrarily large q, and hence N, satisfying Lemma 11.1 is ensured by Lemma 2.1 and the irrationality of λ_1/λ_2. Such N may occur rather infrequently, but at any rate there are infinitely many of them. The lemma fails if no specialization of this kind is made. For instance it can be shown that, for suitable λ_1/λ_2,

$$\limsup_{N \to \infty} \left(\frac{1}{N} \sup_{\mathfrak{m}} \min(|f_1(\alpha)|, |f_2(\alpha)|) \right) > 0.$$

Proof of Lemma 11.1 Suppose that $N \geqslant N_0(\lambda_1, \ldots, \lambda_s)$, let $\alpha \in \mathfrak{m}$ and $Q = N^{k - \nu/2}$ and choose q_j, a_j, in accordance with Lemma 2.1, so that

$$(q_j, a_j) = 1, \quad q_j \leqslant Q, \quad |\lambda_j \alpha - a_j/q_j| \leqslant 1/(q_j Q).$$

The first step is to show that at least one of q_1, q_2 is relatively large. If a_j were to be 0, then one would have

$$|\alpha| \leqslant 1/(q_j Q |\lambda_j|) < N^{\nu - k}$$

and so α would lie in $\mathfrak{M}$, not $\mathfrak{m}$. Thus $a_j \neq 0$. Also one has

$$\lambda_j \alpha = \frac{a_j}{q_j} + \frac{\theta_j}{q_j Q} = \frac{a_j}{q_j}\left(1 + \frac{\theta_j}{a_j Q}\right) \qquad \text{with } |\theta_j| \leqslant 1.$$

Hence

$$\frac{\lambda_1}{\lambda_2} = \frac{\lambda_1 \alpha}{\lambda_2 \alpha} = \frac{q_2 a_1}{a_2 q_1}\left(1 + \frac{\theta_1}{a_1 Q}\right)\left(1 + \frac{\theta_2}{a_2 Q}\right)^{-1}$$

which, since N, and so Q, is large, gives

$$\tfrac{1}{2}\left|\frac{\lambda_1}{\lambda_2}\right| < \left|\frac{q_2 a_1}{a_2 q_1}\right| < 2\left|\frac{\lambda_1}{\lambda_2}\right|$$

and therefore

$$\frac{\lambda_1}{\lambda_2} = \frac{q_2 a_1}{a_2 q_1} + O(Q^{-1}).$$

On hypothesis,

$$\frac{\lambda_1}{\lambda_2} = \frac{a}{q} + \theta q^{-2} \quad \text{with} \quad |\theta| \leqslant 1.$$

Hence

$$\frac{a}{q} - \frac{q_2 a_1}{a_2 q_1} \ll Q^{-1} + q^{-2} \ll N^{-1} = q^{-2},$$

so that

$$|a_2 q_1 a - q_2 a_1 q| \ll |a_2 q_1|/q.$$

If the left-hand side is non-zero, then $|a_2 q_1| \gg q$, and if it is zero, then $a/q = (q_2 a_1)/(a_2 q_1)$, which again implies that $|a_2 q_1| \gg q$. Since $a_2 = \lambda_2 \alpha q_2 - \theta_2 Q^{-1} \ll q_2 P$ it follows that $q_1 q_2 \gg q P^{-1}$. Therefore, by (11.7),

$$\max(q_1, q_2) > N^{1/5}. \tag{11.13}$$

Now, by Weyl's inequality (Lemma 2.4), for $j = 1, 2$,

$$\begin{aligned} f_j(\alpha) &\ll N^{1+\varepsilon}\left(\frac{1}{q_j} + \frac{1}{N} + \frac{q_j}{N^k}\right)^{2^{1-k}} \\ &\ll N^{1+\varepsilon} q_j^{-2^{1-k}} + N^{1-\delta}. \end{aligned}$$

Hence, by (11.13),

$$\min(|f_1(\alpha)|, |f_2(\alpha)|) \ll N^{1-\delta}$$

as required

For the remainder of the proof of Theorem 11.2 it will be assumed that N is chosen in accordance with the specialization given in Lemma 11.1. Let

$$\mathfrak{m}_1 = \{\alpha : \alpha \in \mathfrak{m}, |f_1(\alpha)| \leqslant |f_2(\alpha)|\}, \quad \mathfrak{m}_2 = \mathfrak{m} \setminus \mathfrak{m}_1.$$

By (11.3), $K(\alpha) \ll \min(1, \alpha^{-2})$. Also the argument that gives (11.11) can be readily adapted to show that

$$\int_X^{X+1} \left| \prod_{\substack{i=1 \\ i \neq j}}^{2^k+1} f_i(\alpha) \right| \mathrm{d}\alpha \ll N^{2^k - k + \varepsilon},$$

so that

$$\int_{\mathfrak{m}} \left| \prod_{\substack{i=1 \\ i \neq j}}^{2^k+1} f_i(\alpha) \right| K(\alpha) \mathrm{d}\alpha \ll N^{2^k - k + \varepsilon}.$$

Therefore, by Lemma 11.1, when $j = 1$ or 2,

$$\int_{\mathfrak{m}_j} \left| \prod_{i=1}^{2^k+1} f_i(\alpha) \right| K(\alpha) \mathrm{d}\alpha \ll N^{2^k+1-k-\delta+\varepsilon}.$$

Thus, there is a positive number δ such that

$$\int_{\mathfrak{m}} \left| \prod_{j=1}^{s} f_j(\alpha) \right| K(\alpha) \mathrm{d}\alpha \ll N^{s-k-\delta}. \tag{11.14}$$

11.4 The major arc

In view of (11.12) and (11.14) it remains only to show that, for N sufficiently large,

$$\int_{\mathfrak{M}} \left(\prod_{j=1}^{s} f_j(\alpha) \right) K(\alpha) \mathrm{d}\alpha \gg N^{s-k}. \tag{11.15}$$

Let $\alpha \in \mathfrak{M}$. By (11.7), (11.8), Lemma 2.7, and the remark after the proof of that lemma, one has

$$f_j(\alpha) = v_j(\alpha) + O(N^{2\nu})$$

where

$$v_j(\alpha) = \int_0^N e(\lambda_j \alpha \beta^k) \mathrm{d}\beta. \tag{11.16}$$

Therefore

$$\begin{aligned} f_1 \ldots f_s - v_1 \ldots v_s &= \sum_{j=1}^{s} (f_j - v_j) \left(\prod_{i<j} f_i \right) \left(\prod_{i>j} v_i \right) \\ &\ll N^{s-1+2\nu}. \end{aligned}$$

Hence, by (11.8),

$$\int_{\mathfrak{M}} \left(\prod_{j=1}^{s} f_j(\alpha) - \prod_{j=1}^{s} v_j(\alpha) \right) K(\alpha) \mathrm{d}\alpha \ll N^{s-k-\delta}. \tag{11.17}$$

A change of variables in the integral in (11.16), together with the observation that

$$\int_0^X \frac{1}{k} \gamma^{1/k-1} e(\gamma) \mathrm{d}\gamma$$

is $\ll 1$ uniformly in $X \geqslant 0$, shows that

$$v_j(\alpha) \ll |\alpha|^{-1/k}.$$

Hence, by (11.7) and (11.8),

$$\int_{\mathbb{R}\setminus\mathfrak{M}}\left(\prod_{j=1}^{s} v_j(\alpha)\right)K(\alpha)\mathrm{d}\alpha \ll \int_{N^{\nu-k}}^{\infty} \alpha^{-s/k}\mathrm{d}\alpha$$
$$\ll N^{(s-k)(1-\nu/k)}.$$

Thus

$$\int_{\mathfrak{M}}\left(\prod_{j=1}^{s} v_j(\alpha)\right)K(\alpha)\mathrm{d}\alpha = \int_{-\infty}^{\infty}\left(\prod_{j=1}^{s} v_j(\alpha)\right)K(\alpha)\mathrm{d}\alpha$$
$$+ O(N^{s-k-\delta}). \tag{11.18}$$

By (11.16),

$$\int_{-\infty}^{\infty}\left(\prod_{j=1}^{s} v_j(\alpha)\right)K(\alpha)\mathrm{d}\alpha$$
$$= \int_{-\infty}^{\infty}\mathrm{d}\alpha \int_0^N \mathrm{d}\beta_1 \ldots \int_0^N e((\lambda_1\beta_1^k + \ldots + \lambda_s\beta_s^k)\alpha)K(\alpha)\mathrm{d}\beta_s.$$

Since $K(\alpha) \ll \min(1, \alpha^{-2})$ and the integrand is continuous the order of integration may be interchanged. Hence, by (11.3) and (11.4),

$$\int_{-\infty}^{\infty}\left(\prod_{j=1}^{s} v_j(\alpha)\right)K(\alpha)\mathrm{d}\alpha$$
$$= \int_0^N \mathrm{d}\beta_1 \ldots \int_0^N \max(1 - |\lambda_1\beta_1^k + \ldots + \lambda_s\beta_s^k|, 0)\mathrm{d}\beta_s$$
$$= k^{-s}\int_0^{N^k} \mathrm{d}\alpha_1 \ldots \int_0^{N^k} (\alpha_1 \ldots \alpha_s)^{1/k-1}$$
$$\times \max(1 - |\lambda_1\alpha_1 + \ldots + \lambda_s\alpha_s|, 0)\mathrm{d}\alpha_s.$$

It is now that one requires the hypothesis that $\lambda_1/\lambda_2 < 0$. Consider the region

$$\mathscr{B} = \{(\alpha_2, \ldots \alpha_s) : \delta N^k \leqslant \alpha_2 \leqslant 2\delta N^k, \delta^2 N^k \leqslant \alpha_j \leqslant 2\delta^2 N^k\ (3 \leqslant j \leqslant s)\}.$$

Then, for δ sufficiently small, whenever $(\alpha_2, \ldots, \alpha_s) \in \mathscr{B}$ one has

$$2\delta^2 N^k < -(\lambda_2\alpha_2 + \ldots + \lambda_s\alpha_s)\lambda_1^{-1} < \tfrac{1}{2}N^k$$

and so every α_1 with $|\lambda_1\alpha_1 + \ldots + \lambda_s\alpha_s| \leqslant \frac{1}{2}$ satisfies $\delta^2 N^k < \alpha_1 < N^k$. Therefore

$$\int_{-\infty}^{\infty}\left(\prod_{j=1}^{s} v_j(\alpha)\right)K(\alpha)\mathrm{d}\alpha \gg (N^{1-k})^s \int_{\mathscr{B}} \mathrm{d}\alpha_2 \ldots \mathrm{d}\alpha_s \int_{\mathscr{A}(\alpha_2, \ldots, \alpha_s)} \mathrm{d}\alpha_1$$

where $\mathscr{A}(\alpha_2, \ldots, \alpha_s)$ denotes the interval with end points $(-(\lambda_2\alpha_2 + \ldots + \lambda_s\alpha_s) \pm \frac{1}{2})\lambda_1^{-1}$. Obviously the volume of $\mathscr{B}$ is $\gg (N^k)^{s-1}$. Hence

$$\int_{-\infty}^{\infty} \left(\prod_{j=1}^{s} v_j(\alpha) \right) K(\alpha) \mathrm{d}\alpha \gg N^{s-k}.$$

This with (11.17) and (11.18) establishes (11.15), and thus completes the proof of Theorem 11.2.

11.5 Exercises

1 (Davenport & Roth, 1955; Vaughan, 1974*b*) Obtain Theorem 11.1 for any $s \geqslant Ck \log k$ where C is a suitable constant.

2 Let $\lambda_1, \lambda_2, \lambda_3, \mu, \eta$ denote real numbers with $\lambda_j \neq 0$, $\eta > 0$, λ_1/λ_2 irrational, and $\lambda_1/\lambda_2 < 0$. Show that there are primes p_1, p_2, p_3 such that

$$|\lambda_1 p_1 + \lambda_2 p_2 + \lambda_3 p_3 + \mu| < \eta.$$

3 (Baker, 1967; Vaughan, 1974*a*) Modify the argument used to answer 2 above so as to show that there are infinitely many triples of primes p_1, p_2, p_3 such that

$$|\lambda_1 p_1 + \lambda_2 p_2 + \lambda_3 p_3 + \mu| < (\log \max_j p_j)^{-\eta}.$$

4 (Baker).† Let $F(N) \to 0$ as $N \to \infty$. Prove that the statement 'for every sufficiently large N there are primes p_1, p_2, p_3 such that $p_j \leqslant N$ and $|\lambda_1 p_1 + \lambda_2 p_2 + \lambda_3 p_3| > F(N)$' can be false for suitable $\lambda_1, \lambda_2, \lambda_3$ with $\lambda_1/\lambda_2 > 0$ and λ_1/λ_2 irrational.

†Communicated in conversation in June 1973.

Bibliography

Works are listed here alphabetically by author(s). Those by the same author(s) are listed chronologically.

Numbers 1, . . . , 11 or the letters B, E, G, S have been added in square brackets at the end of each entry to indicate either that the work is related to the corresponding chapter or that it is

B. Basic material,

E. An Exposition or monograph covering some aspects of the Hardy–Littlewood method,

G. A Generalization or development of the Hardy–Littlewood method,

S. A Survey article.

Apostol, T. M. (1976). *Introduction to analytic number theory*. New York: Springer Verlag. [B].

Arhipov, G. I. & Karatsuba, A. A. (1978). A new estimate of an integral of I. M. Vinogradov. *Izv. Akad. Nauk SSSR, ser. mat.*, **42**, 751–62. [5].

Arkhangelskaya, V. M. (1957). Some calculations connected with Goldbach's problem. *Ukraine Math. J.*, **9**, 20–9. [3].

Ayoub, R. (1953*a*). On Rademacher's extension of the Goldbach–Vinogradoff theorem. *Trans. Am. Math. Soc.*, **74**, 482–91. [G].

Ayoub, R. (1953*b*). On the Waring–Siegel theorem. *Can. J. Math.*, **5**, 439–50. [G].

Babaev, G. & Subhankulov, M. A. (1963). An asymptotic formula for two additive problems. *Tadjhik. Gos. Univ. Utsen. Zap.*, **26**, 49–68. [G].

Baker, A. (1967). On some diophantine inequalities involving primes. *J. Reine Angew. Math.*, **228**, 166–81. [11].

Bambah, R. P. (1954). Four squares and a *k*-th power. *Q. J. Math.*, **5**, 191–202. [11].

Batchelder, P. M. (1936). Waring' problem. *Am. Math. Month.*, **43**, 21–7. [1, S].

Behrend, F. A. (1946). On sets of integers which contain no three terms in arithmetical progression. *Proc. Natn. Acad. Sci. U.S.A.*, **32**, 331–2. [10].

Bierstedt, R. G. (1963). Some problems on the distribution of *k*th power residues modulo a prime. Ph.D. thesis. University of Colorado, Boulder. [9].

Birch, B. J. (1957). Homogeneous forms of odd degree in a large number of variables. *Mathematika*, **4**, 102–5. [9].

Birch, B. J. (1962). Forms in many variables. *Proc. R. Soc. Lond.*, **265A**, 245–63.

Birch, B. J. (1970). Small zeros of diagonal forms of odd degree in many variables. *Proc. Lond. Math. Soc.*, (3), **21**, 12–18. [9].

Birch, B. J. & Davenport, H. (1958). On a theorem of Davenport and Heilbronn. *Acta Math.*, **100**, 259–79. [11].

Birch, B. J., Davenport, H. & Lewis, D. J. (1962). The addition of norm forms. *Mathematika*, **9**, 75–82. [G].

Bombieri, E. & Davenport, H. (1966). Small differences between prime numbers. *Proc. R. Soc. Lond.*, **293A**, 1–18. [G.].

Bovey, J. D. (1974). Γ*(8). *Acta Arith.*, **25**, 145–50. [9].

Brauer, R. (1945). A note on systems of homogeneous algebraic equations. *Bull. Am. Math. Soc.*, **51**, 749–55. [9].

Cassels, J. W. S. (1960). On the representation of integers as the sums of distinct summands taken from a fixed set. *Acta Sci. Math. Szeged*, **21**, 111–24. [10].

Cauchy, A. L. (1813). Recherches sur les nombres. *J. Ec. Polytech.*, **9**, 99–116. [2].

Chen, J. -R. (1958). On Waring's problem for n-th powers. *Acta Math. Sinica*, **8**, 253–7, translated in *Chin. Math. Acta*, **8** (1966), 849–53. [5].

Chen, J. -R. (1959). On the representation of a natural number as a sum of terms of the form $x(x+1)\ldots(x+k-1)/k!$. *Acta Math. Sinica.*, **9**, 264–70. [G].

Chen, J. -R. (1964). Waring's problem for $g(5)=37$. *Scientia Sinica*, **13**, 335 and 1547–68. See also *Sci. Rec.*, **3** (1959), 327–30. [1].

Chen, J. -R. (1965). On large odd numbers as sums of three almost equal primes. *Scientia Sinica*, **14**, 1113–17. [3].

Chowla, I. (1935*a*). A theorem on the addition of residue classes. *Proc. Indian Acad. Sci.*, **2**, 242–3. [2].

Chowla, I. (1935*b*). A theorem on the addition of residue classes: Application to the number $\Gamma(k)$ in Waring's problem. *Proc. Indian Math. Soc.*, **2A**, 242–3, and *Q. J. Math.*, **8** (1937), 99–102. [4].

Chowla, I. (1937*a*). On $\Gamma(k)$ in Waring's problem and an analogous function. *Proc. Indian Acad. Sci.*, **5A**, 269–76. [4].

Chowla, I. (1937*b*). A new evaluation of the number $\Gamma(k)$ in Waring's problem. *Proc. Indian Acad. Sci.*, **6A**, 97–103. [4].

Chowla, S. D. (1934). A theorem on irrational indefinite quadratic forms. *J. Lond. Math. Soc.*, **9**, 162–3. [11].

Chowla, S. D. (1936). Pillai's exact formula for the number $g(n)$ in Waring's problem. *Proc. Indian Acad. Sci.*, **3A**, 339–40 and **4**, 261. [1].

Chowla, S. D. (1944). On $g(k)$ in Waring's problem. *Proc. Lahore Philos. Soc.*, **6**, 16–17. [1].

Chowla, S. D. (1960). On a conjecture of J. F. Gray, *Norske Vid. Selsk. Forh. (Trondheim)*, **33**, 58–9. [9].

Chowla, S. D. (1961). On the congruence $\Sigma_{i=1}^{s} a_i x_i^k \equiv 0 \pmod p$, *J. Indian Math. Soc.*, **25**, 47–8. [9].

Chowla, S. D. (1963). On a conjecture of Artin, I, II. *Norske Vid. Selsk. Forh. (Trondheim)*, **36**, 135–41. [9].

Chowla, S. D. & Davenport, H. (1960/1961). On Weyl's inequality and Waring's problem for cubes. *Acta Arith.*, **6**, 505–21. [9].

Chowla, S. D. & Shimura, G. (1963). On the representation of zero by a linear combination of k-th powers, *Norske Vid. Selsk. Forh. (Trondheim)*, **36**, 169–76. [9].

Chudakov, N. G. (1937). On the Goldbach problem. *C. R. Acad. Sci. URSS*, (2), **17**, 335–8.

Chudakov, N. G. (1938). On the density of the set of even numbers which are not representable as a sum of two odd primes. *Izv. Akad. Nauk SSSR Ser. Nat.*, **2**, 25–40. [3].

Chudakov, N. G. (1947). On the Goldbach–Vinogradov's theorem. *Ann. Math.*, (2), **48**, 515–45. [3].

Cook, R. J. (1971). Simultaneous quadratic equations. *J. Lond. Math. Soc.*, (2), **4**, 319–26. [G].

Cook, R. J. (1972*a*). A note on a lemma of Hua. *Q. J. Math.*, **23**, 287–8. [G].

Cook, R. J. (1972*b*). Pairs of additive equations. *Michigan Math. J.*, **19**, 325–31. [G].

Cook, R. J. (1973*a*). A note on Waring's problem. *Bull. Lond. Math. Soc.*, **5**, 11–12. [6].

Cook, R. J. (1973*b*). Simultaneous quadratic equations II. *Acta Arith.*, **25**, 1–5. [G].

Cook, R. J. (1974). Simultaneous quadratic inequalities. *Acta Arith.*, **25**, 337–46. [G].

Cook, R. J. (1975). Indefinite hermitian forms. *J. Lond. Math. Soc.*, (2), **11**, 107–12. [G].

Cook, R. J. (1977, 1979). Diophantine inequalities with mixed powers I, II. *J. Number Theor.*, **9**, 261–74; **11**, 49–68. [G].

Corput, J. G. van der (1937a). Sur le théorème de Goldbach–Vinogradov, *C. R. Acad. Sci., Paris*, **205**, 479–81. [3].

Corput, J. G. van der (1937*b*). Une nouvelle généralisation du théorème de Goldbach–Vinogradov. *C. R. Acad. Sci. Paris*, **205**, 591–2. [3].

Corput, J. G. van der (1937*c*). Sur l'hypothèse de Goldbach pour presque tous les nombres pairs. *Acta Arith.*, **2**, 266–90. [3].

Corput, J. G. van der (1937*d*, 1938*a,b,c,d*). Sur deux, trois ou quatre nombres premiers, I, II, III, IV, V. *Proc. Akad. Wet. Amsterdam*, **40**, 846–51; **41**, 25–36, 97–107, 217–26, 344–49. [G].

Corput, J. G. van der (1938*e*). Sur l'hypothèse de Goldbach. *Proc. Akad. Wet. Amsterdam*, **41**, 76–80. [3].

Corput, J. G. van der (1938*f*). Uber Summen von Primzahlen und Primzahlen quadraten. *Math. Ann.*, **116**, 1–50. [G].

Corput, J. G. van der (1938*g,h,i,j*, 1939). Contribution à la théorie additive des nombres I, II, III, IV, V. *Proc. Akad. Wet. Amsterdam*, **41**, 227–37, 350–61, 442–53, 556–67; **42**, 336–45. [G].

Corput, J. G. van der & Pisot, Ch. (1939). Sur un problème de Waring généralisé III, *Proc. Akad. Wet. Amsterdam*, **42**, 566–72. [G].

Danicic, I. (1958). The solubility of certain Diophantine inequalities. *Proc. Lond. Math. Soc.*, (3), **8**, 161–76. [11].

Danicic, I. (1966). On the integral part of a linear form with prime variables. *Can. J. Math.*, **18**, 621–28. [11].

Davenport, H. (1935). On the addition of residue classes. *J. Lond Math. Soc.*, **10**, 30–2. [2].

Davenport, H. (1938). Sur les sommes de puissances entières. *C. R. Acad. Sci.*, Paris, **207**, 1366–8. [6].

Davenport, H. (1939*a*). On Waring's problem for cubes. *Acta Math.*, **71**, 123–43. [6].

Davenport, H. (1939*b*). On sums of positive integral *k*th powers. *Proc. R. Soc. Lond.*, **170A**, 293–9. [6].

Davenport, H. (1939*c*). On Waring's problem for fourth powers. *Ann. Math.*, **40**, 731–47. [6].

Davenport, H. (1942*a*). On sums of positive integral *k*th powers. *Am. J. Math.*, **64**, 189–98. [6].

Davenport, H. (1942*b*). On Waring's problem for fifth and sixth powers. *Am. J. Math.*, **64**, 199–207. [6].

Davenport, H. (1947). A historical note. *J. Lond. Math. Soc.*, **22**, 100–1. [2].

Davenport, H. (1950). Sums of three positive cubes. *J. Lond. Math. Soc.*, **25**, 339–43. [6].

Davenport, H. (1956, 1958). Indefinite quadratic forms in many variables I, II. *Mathematika*, **3**, 81–101; *Proc. Lond. Math. Soc.*, (3), **8**, 109–26. [11].

Davenport, H. (1959). Cubic forms in thirty two variables. *Philos. Trans. R. Soc. Lond.*, **261A**, 193 – 210. [9].

Davenport, H. (1960*a*). Über einige neuere Fortschritte der additiven Zahlentheorie. *Jahresbr. der Deutschen Math. Ver.*, **63**, 163–9. [S].

Davenport, H. (1960*b*). Some recent progress in analytic number theory. *J. Lond. Math. Soc.*, **35**, 135–42. [S].

Davenport, H. (1962*a*). Cubic forms in 29 variables. *Proc. R. Soc. Lond.*, **266A**, 287–98. [9].

Davenport, H. (1962*b*). *Analytic methods for Diophantine equations and Diophantine inequalities.* Ann Arbor: Ann Arbor Publishers. [E].

Davenport, H. (1963). Cubic forms in sixteen variables. *Proc. R. Soc. Lond.*, **272A**, 285–303. [9].

Davenport, H. (1966). *Multiplicative number theory.* 1st edn. Chicago: Markham. 2nd ed. revised by Montgomery, H. L. (1980). Graduate Texts in Mathematics, 74. Berlin: Springer–Verlag. [B].

Davenport, H. (1977). *The collected works of Harold Davenport*, vol. III, ed. B. J. Birch, H. Halberstram & C. A. Rogers. London: Academic Press. [G].

Davenport, H. & Erdős, P. (1939). On sums of positive integral kth powers. *Ann. Math.*, **40**, 533–6. [6].

Davenport, H. & Heilbronn, H. (1936*a*). On Waring's problem for fourth powers. *Proc. Lond. Math. Soc.*, (2), **41**, 143–50. [5].

Davenport, H. & Heilbronn, H. (1936*b*). On an exponential sum. *Proc. Lond. Math. Soc.*, (2), **41**, 449–53 [4].

Davenport, H. & Heilbronn, H. (1937*a*). On Waring's problem: two cubes and one square. *Proc. Lond. Math. Soc.*, (2), **43**, 73–104. [8].

Davenport, H. & Heilbronn, H. (1937*b*). Note on a result in the additive theory of numbers. *Proc. Lond. Math. Soc.*, (2), **43**, 142–51. [G].

Davenport, H. & Heilbronn, H. (1946). On indefinite quadratic forms in five variables. *J. Lond. Math. Soc.*, **21**, 185–93. [11].

Davenport, H. & Lewis, D. J. (1963). Homogeneous additive equations. *Proc. R. Soc. Lond.*, **274A**, 443–60. [9].

Davenport, H. & Lewis, D. J. (1966). Cubic equations of additive type. *Philos. Trans. R. Soc. Lond.*, **261A**, 97–136. [G].

Davenport, H. & Lewis, D. J. (1969*a*). Simultaneous equations of additive type. *Philos. Trans. R. Soc. Lond.*, **264A**, 557–95. [G].

Davenport, H. & Lewis, D. J. (1969*b*). Two additive equations. *American Mathematical Society Proceedings of Symposia in Pure Mathematics*, **12**, 74–98. [G].

Davenport, H. & Lewis, D. J. (1972). Gaps between values of positive definite quadratic forms. *Acta Arith.*, **21**, 87–105. [G].

Davenport, H. & Ridout, D. (1959). Indefinite quadratic forms. *Proc. Lond. Math. Soc.*, (3), **9**, 544–55. [G].

Davenport, H. & Roth, K. F. (1955). The solubility of certain Diophantine inequalities. *Mathematika*, **2**, 81–96. [11].

Dickson, L. E. (1933). Recent progress on Waring's theorem and its generalizations. *Bull. Am. Math. Soc.*, **39**, 701–27. [1].

Dickson, L. E. (1936*a*). *Researches on Waring's problem.* Carnegie Inst. of Washington Publ. **464**. [1].

Dickson, L. E. (1936*b*). Proof of the ideal Waring theorem for exponents 7–180. *Am. J. Math.*, **58**, 521–9. [1].

Dickson, L. E. (1936*c*). Solution of Waring's problem. *Am. J. Math.*, **58**, 530–5. [1].

Dickson, L. E. (1936*d*). The Waring problem and its generalizations. *Bull. Am. Math. Soc.*, **42**, 833–42. [1].

Dickson, L. E. (1936*e*). On Waring's problem and its generalization. *Ann. Math.*, **37**, 293–316. [1].

Dickson, L. E. (1936*f*). The ideal Waring theorem for twelfth powers. *Duke Math. J.*, **2**, 192–204. [1].

Dickson, L. E. (1936*g*). Universal Waring theorems. *Monatshefte für Mathematik und Physik*, **43**, 391–400. [1].

Dodson, M. M. (1967). Homogeneous additive congruences. *Philos. Trans. R. Soc. Lond.*, **261A**, 163–210. [9].

Ehlich, H. (1965). Zur Pillaischen Vermutung. *Arch. Math.*, **16**, 223–26. [1].

Ellison, W. J. (1971). Waring's problem. *Am. Math. Mon.*, **78**, 10–36. [1].

Emel'yanov, G. V. (1950). On a system of Diophantine equations. *Leningrad Gos. Univ. Uch. Zap.* **137**, *Ser. Mat. Nauk*, **19**, 3–39. [G].

Erdős, P. & Turán, P. (1936). On some sequences of integers. *J. Lond. Math. Soc.*, **11**, 261–4. [10].

Erdős, P. & Vaughan, R. C. (1974). Bounds for the *r*th coefficients of cyelotomic polynomials. *J. Lond. Math. Soc.*, (2), **8**, 393–400. [3].

Estermann, T. (1929). On the representation of a number as the sum of three products. *Proc. Lond. Math. Soc.*, (2), **29**, 453–78. [G].

Estermann, T. (1929). Vereinfachter Beweis eines Satzes von Kloosterman. *Abhandlungen aus dem Mathematischen Seminar der Hamburgischen Universität*, **7**, 82–98. [G].

Estermann, T. (1930*a,b*). On the representation of a number as the sum of two products, I, II. *Proc. Lond. Math. Soc.*, (2), **31**, 123–133; *J. Lond. Math. Soc.*, **5**, 131–7. [G].

Estermann, T. (1936). Proof that every large integer is a sum of seventeen biquadrates. *Proc. Lond. Math. Soc.*, (2), **41**, 126–42. [6].

Estermann, T. (1937*a*). On Waring's problem for fourth and higher powers. *Acta Arith.*, **2**, 197–211. [5].

Estermann, T. (1937*b*). Proof that every large integer is the sum of two primes and a square. *Proc. Lond. Math. Soc.*, (2), **42**, 501–16. [G].

Estermann, T. (1937*c*). A new result in the additive prime number theory. *Q. J. Math.*, **8**, 32–8. [3].

Estermann, T. (1938). On Goldbach's problem: Proof that almost all even positive integers are sums of two primes. *Proc. Lond. Math. Soc.*, (2), **44**, 307–14. [3].

Estermann, T. (1948). On Waring's problem: A simple proof of a theorem of Hua. *Sci. Rep. Natn. Tsing Hua Univ.*, **5A**, 226–39. [2].

Estermann, T. (1951). On sums of squares of square-free numbers. *Proc. Lond. Math. Soc.*, (2), **53**, 125–37. [G].

Estermann, T. (1952). *Introduction to modern prime number theory.* Cambridge University Press. [E].

Estermann, T. (1962). A new application of the Hardy–Littlewood–Kloosterman method. *Proc. Lond. Math. Soc.*, (3), **12**, 425–44. [G].

Evelyn, C. J. A. & Linfoot, E. H. (1929, 1933). On a problem in the additive theory of numbers I, VI, *Math. Z.*, **30**, 433–48; *Q. J. Math.*, **4**, 309–14. [G].

Földes, I. (1952). On the Goldbach hypothesis concerning the prime numbers of an arithmetical progression. *C. R. Prem. Cong. Mat. Hongrois*, 473–92. [3].

Fowler, J. (1962). A note on cubic equations. *Proc. Camb. Philos. Soc.*, **58**, 165–69. [9].

Freiman, G. A. (1949). Solution of Waring's problem in a new form. *Uspehi Mat. Nauk*, **4**, 193. [5,8].

Furstenburg, H. (1977). Ergodic behaviour of diagonal measures and a theorem of Szemerédi on arithmetic progressions. *J. d'Analyse Math.*, **31**, 204–56. [10].

Gallagher, P. X. (1975). Primes and powers of 2. *Inventiones Math.*, **29**, 125–42. [G].

Gelbcke, M. (1931). Zum Waringschen Problem. *Math. Ann.*, **105**, 637–52. [2].

Gelbcke, M. (1933). A propos de $g(k)$ dans le problème de Waring. *C. R. Acad. Sci. URSS*, (7), 631–40. [2].

Ghosh, A. (1981). The distribution of αp^2 modulo one. *Proc. Lond. Math. Soc.*, **42**, [G].

Gray, J. F. (1960). Diagonal forms of odd degree over a finite field. *Michigan Math. J.*, **7**, 297–301. [9].

Grosswald, E. (1968/9). On some conjectures of Hardy and Littlewood. *Publ. Ramanujan Inst.*, **1**, 75–89. [8].

Halberstam, H. (1950). Representation of integers as sums of a square, a positive cube, and a fourth power of a prime. *J. Lond. Math. Soc.*, **25**, 158–68. [G].

Halberstam, H. (1951*a*). Representation of integers as sums of a square of a prime, a cube of a prime, and a cube. *Proc. Lond. Math. Soc.* (2), **52**, 455–66. [G].

Halberstam, H. (1951*b*). On the representation of large numbers as sums of squares, higher powers, and primes. *Proc. Lond. Math. Soc.*, (2), **53**, 363–80. [G].

Halberstam, H. (1957). An asymptotic formula in the theory of numbers. *Trans. Am. Math. Soc.*, **84**, 338–51. [G].

Hardy, G. H. (1922). Goldbach's theorem. *Math. Tid. B*, 1–16. [1].

Hardy, G. H. (1966). *Collected papers of G. H. Hardy, including joint papers with J. E. Littlewood and others*, ed. by a committee appointed by the London Mathematical Society, vol. I. Oxford: Clarendon Press. [E].

Hardy, G. H. & Littlewood, J. E. (1919). A new solution of Waring's problem. *Q. J. Math.*, **48**, 272–93. [1,2].

Hardy, G. H. & Littlewood, J. E. (1920). Some problems of "Partitio Numerorum". I A new solution of Waring's problem. *Göttingen Nachrichten*, 33–54, [1, 2].

Hardy, G. H. & Littlewood, J. E. (1921). Some problems of "Partitio Numerorum": II Proof that every large number is the sum of at most 21 biquadrates. *Math. Z.*, **9**, 14–27. [1,6].

Hardy, G. H. & Littlewood, J. E. (1922). Some problems of "Partitio Numerorum": IV The singular series in Waring's problem. *Math. Z.*, **12**, 161–88. [4].

Hardy, G. H. & Littlewood, J. E. (1923*a*). Some problems of "Partitio Numerorum": III On the expression of a number as a sum of primes. *Acta Math.*, **44**, 1–70. [1,3].

Hardy, G. H. & Littlewood, J. E. (1923*b*). Some problems of "Partitio Numerorum": V A further contribution to the study of Goldbach's problem. *Proc. Lond. Math. Soc.*, (2), **22**, 46–56. [1,3].

Hardy, G. H. & Littlewood, J. E. (1925). Some problems of "Partitio Numerorum": VI Further researches in Waring's problem. *Math. Z.*, **23**, 1–37. [4, 6].

Hardy, G. H. & Littlewood, J. E. (1928). Some problems of "Partitio Numerorum": VIII† The number $\Gamma(k)$ in Waring's problem. *Proc. Lond. Math. Soc.*, (2), **28**, 518–42. [4].

Hardy, G. H., Littlewood, J. E. & Pólya, G. (1951). *Inequalities*, 2nd edn. Cambridge University Press. [B].

Hardy, G. H. & Ramanujan, S. (1918). Asymptotic formulae in combinatory analysis. *Proc. Lond. Math. Soc.*, (2), **17**, 75–115. [1].

Hardy, G. H. & Wright, E. M. (1979). *An introduction to the theory of numbers*, 5th edn. Oxford: Oxford University Press. [B].

Hasse, H. (1964). *Vorlesungen über Zahlentheorie. Zweite auflage.* Berlin: Springer-Verlag. [B].

Heilbronn, H. (1936). Über das Waringsche Problem. *Acta Arith.*, **1**, 212–21. [5].

Hilbert, D. (1909*a*,*b*). Beweis für die Darstellbarkeit der ganzen Zahlen durch eine feste Anzahl nter Potenzen (Waringsche Problem). Nachrichten von der Königlichen Gesellchaft der Wissenschaften zu Göttingen, mathematisch-physikalische Klasse aus den Jahren 1909, 17–36; *Math. Annalen*, **67**, 281–300. [1].

Householder, J. E. (1959). The representation of zero by odd *k*th power diagonal forms. Ph.D. Thesis. University of Colorado, Boulder. [9].

Hua, L. -K. (1935). On Waring theorems with cubic polynomial summands. *Math. Ann.*, **111**, 622–8. [G].

Hua, L. -K. (1936*a*,*b*). On Waring's problem with polynomial summands. *Am. J. Math.*, **58**, 553–62; *J. Chin. Math. Soc.*, **1**, 21–61. [G].

Hua, L. -K. (1937*a*). On a generalized Waring problem. *Proc. Lond. Math. Soc.*, (2), **43**, 161–82.

Hua, L. -K. (1937*b*). On the representation of integers as the sums of *k*th powers of primes. *C. R. Acad. Sci. URSS*, (2), **17**, 167–8. [G].

Hua, L. -K. (1938*a*). Some results on Waring's problem for smaller powers. *C. R. Acad. Sci. URSS*, (2), **18**, 527–8. [6].

Hua, L. -K. (1938*b*). On Waring's problem. *Q. J. Math.*, **9**, 199–202. [2].

Hua, L. -K. (1938*c*,*d*). Some results in the additive prime number theory. *C. R. Acad. Sci. URSS*, (2), **18**, 3; *Q. J. Math.*, **9**, 68–80. [G].

Hua, L.-K. (1939). On Waring's problem for fifth powers. *Proc. Lond. Math. Soc.*, (2), **45**, 144–60. [6].

Hua, L.-K. (1940*a*). Sur une somme exponentielle. *C. R. Acad. Sci. Paris*, **210**, 520–3. [7].

Hua, L. -K. (1940*b*). Sur le problème de Waring relatif à un polynome du troisième degré. *C. R. Acad. Sci. Paris*, **210**, 650–2. [G].

Hua, L.-K. (1940*c*). On a system of Diophantine equations. *Dokl. Akad. Nauk SSSR*, **27**, 312–13. [G].

Hua, L. -K. (1940*d*). On a generalized Waring problem II. *J. Chin. Math. Soc.*, **2**, 175–91. [G].

Hua, L. -K. (1940*e*, *f*). On Waring's problem with cubic polynomial summands. *Sci. Rep. Natn. Tsing Hua Univ.*, **4A**, 55–83; *J. Indian Math. Soc.*, **4**, 127–35. [G].

Hua, L. -K. (1947). Some results on additive theory of numbers. *Proc. Natn. Acad. Sci. U.S.A.*, **33**, 136–7. [G].

† Number VII in this series is an unpublished manuscript on small differences between prime numbers. See Bombieri & Davenport (1966).

Hua, L. -K. (1949). An improvement of Vinogradov's mean value theorem and several applications. *Q. J. Math.*, **20**, 48–61. [5].

Hua, L. -K. (1952). On the number of solutions of Tarry's problem. *Acta Sci. Sinica*, **1**, 1–76. [7].

Hua, L. -K. (1957*a*). On exponential sums. *Sci. Rec.*, **1**, 1–4. [4].

Hua, L. -K. (1957*b*). On the major arcs in Waring's problem. *Sci. Rec.*, **1**, 17–18. [4].

Hua, L. -K. (1959). *Die Abschätzung von Exponentialsummen und ihre anwendung in der Zahlentheorie*. Enzyklopädie der Math. Wiss. Band I, 2. Heft 13. Teil 1. Leipzig: Teubner. [E].

Hua, L. -K. (1965). *Additive theory of prime numbers*. Providence, Rhode Island: American Mathematical Society. [E].

Humphreys, M. G. (1935). On the Waring problem with polynomial summands. *Duke Math. J.*, **1**, 361–75. [G].

Huston, R. E. (1935). Asymptotic generalizations of Waring's theorem. *Proc. Lond. Math. Soc.*, (2), **39**, 82–115. [G].

Huxley, M. N. (1968). The large sieve inequality for algebraic number fields. *Mathematika*, **15**, 178–87. [5].

Huxley, M. N. (1969). On the differences of primes in arithmetical progressions. *Acta Arith.*, **15**, 367–92. [G].

Huxley, M. N. (1973, 1977). Small differences between consecutive primes, I, II. *Mathematika*, **20**, 229–32; **24**, 142–52. [G].

Iseki, K. (1949). A remark on the Goldbach–Vinogradov theorem. *Proc. Jpn. Acad.*, **25**, 185–7. [3].

Iseki, S. (1968). A problem on partitions connected with Waring's problem. *Proc. Am. Math. Soc.*, **19**, 197–204. [2].

James, R. D. (1934*a*). The value of the number $g(k)$ in Waring's problem. *Trans. Am. Math. Soc.*, **36**, 395–444. [2].

James, R. D. (1934*b*). On Waring's problem for odd powers. *Proc. Lond. Math. Soc.*, (2), **37**, 257–91. [2].

James, R. D. & Weyl, H. (1942). Elementary note on prime number problems of Vinogradoff's type. *Am. J. Math.*, **64**, 539–52. [3].

Kalinka, V. (1963). Generalization of a lemma of L. -K. Hua for algebraic numbers. *Litovsk Mat. Sb.*, **3**, 149–55. [G].

Kamke, E. (1921). Verallgemeinerungen des Waring–Hilbertschen Satzes. *Mat. Ann.*, **83**, 85–112. [G].

Kamke, E. (1922). Bemerkung zum allgemein Waringschen Problem. *Mat. Z.*, **15**, 188–94. [G].

Karatsuba, A. A. (1965). On the estimation of the number of solutions of certain equations. *Dokl. Akad. Nauk SSSR*, **165**, 31–2, translated in *Sov. Math. Dokl.*, **6**, 1402–4. [5].

Karatsuba, A. A. (1968). A certain system of indeterminate equations. *Mat. Z.*, **4**, 125–8. [5].

Karatsuba, A. A. & Korobov, N. M. (1963). A mean value theorem. *Dokl. Akad. Nauk SSSR*, **149**, 245–8. [5].

Kestelman, H. (1937). An integral connected with Waring's problem. *J. Lond. Math. Soc.*, **12**, 232–40. [2].

Khintchine, A. (1952). *Three pearls of number theory*. Rochester, N.Y.: Graylock Press. [1].

Kloosterman, H. D. (1925*a*). Over het uitdrukken van geheele positieve getallen in den vorm $ax^2 + by^2 + cz^2 + dt^2$. *Verslag Amsterdam*, **34**, 1011–15. [G].

Kloosterman, H. D. (1925*b*). On the representation of numbers in the form $ax^2 + by^2 + cz^2 + dt^2$. *Acta Math.*, **49**, 407–64. [G].

Kloosterman, H. D. (1925*c*). On the representation of numbers in the form $ax^2 + by^2 + cz^2 + dt^2$. *Proc. Lond. Math. Soc.*, (2), **25**, 143–73. [G].

Körner, O. (1960). Übertragung des Goldbach–Vinogradovschen Satzes auf reell-quadratisch Zahlkörper. *Math. Ann.*, **141**, 343–66. [G].

Körner, O. (1961*a*). Erweiterter Goldbach–Vinogradovscher Satz in beliebigen algebraischen Zahlkörpern. *Math. Ann.*, **143**, 344–78. [G].

Körner, O. (1961*b*). Zur additiven Primzahltheorie algebraischer Zahlkörper. *Math. Ann.*, **144**, 97–109. [G].

Körner, O. (1961*c*). Über das Waringsche Problem in algebraischen Zahlkörper. *Math. Ann.*, **144**, 224–38. [G].

Körner, O. (1962). Über Mittelwerte trigonometrischer Summen und ihre Anwendung in algebraischen Zahlkörpern. *Math. Ann.*, **147**, 205–39, corrections, *ibid*, **149**, (1963), 462. [G].

Körner, O. (1962/3). Ganze algebraische Zahlen als Summen von Polynomwerten. *Math. Ann.*, **149**, 97–104. [G].

Körner, O. (1964). Darstellung ganzer Grössen durch Primzahlpotenzen in algebraischen Zahlkörpern. *Math. Ann.*, **155**, 204–45. [G].

Kovacs, B. (1972). Über die Lösbarkeit diophantischer Gleichungen von additiven Typ. I. *Publ. Math.*, **19**, 259–73. [G].

Landau, E. G. H. (1922). Zur additiven Primzahltheorie. *Palermo Rend.*, **46**, 349–56. [3].

Landau, E. G. H. (1927). *Vorlesungen über Zahlentheorie*. Erster Band. Leipzig: Verlag von S. Hirzel. [E].

Landau, E. G. H. (1930). Über die neue Winogradoffsche Behandlung des Waringschen Problems. *Math. Z.*, **31**, 319–38. [2].

Landau, E. G. H. (1937). *Über einige neuere Fortschritte der additiven Zahlentheorie*. Cambridge University Press. [E].

Lau, K. W. & Liu, M.-C. (1978). Linear approximation by primes. *Bull. Aust. Math. Soc.*, **19**, 457–66. [11].

Lavrik, A. F. (1959). On a theorem in the additive theory of numbers. *Uspehi Mat. Nauk*, **14**, 197–8. [G].

Lavrik, A. F. (1960*a*). On the twin prime hypothesis of the theory of primes by the method of I. M. Vinogradov. *Dokl. Akad. Nauk SSSR*, **132**, 1013–15, translated in *Soviet Math. Dokl.*, **1** (1960), 700–2. [3].

Lavrik, A. F. (1960*b*). On the distribution of k-twin primes. *Dokl. Akad. Nauk SSSR*, **132**, 1258–60, translated in *Soviet Math. Dokl.*, **1** (1960), 764–6. [3].

Lavrik, A. F. (1961*a*). The number of k-twin primes lying in an interval of a given length. *Dokl. Akad. Nauk SSSR*, **136**, 281–3, translated in *Soviet Math. Dokl.*, **2** (1961), 52–5. [3].

Lavrik, A. F. (1961*b*). Binary problems of additive prime number theory connected with the method of trigonometric sums of I. M. Vinogradov. *Vestnik Leningrad Univ.*, **16**, 11–27. [3].

Lavrik, A. F. (1961*c*). On the theory of distribution of primes based on I. M. Vinogradov's method of trigonometric sums. *Trudy Mat. Inst. Steklov*, **64**, 90–125. [3].

Lavrik, A. F. (1961*d*). On the theory of the distribution of sets of primes with given differences between them. *Dokl. Akad. Nauk SSSR*, **138**, 1287–90, translated in *Soviet Math. Dokl.*, **2** (1961), 827–30. [3].

Lavrik, A. F. (1962). On the representation of numbers as the sum of primes by Shnirel'man's method. *Izv. Akad. Nauk UzSSR Ser. Fiz.-Mat. Nauk*, **3**, 5–10. [3].

Lewis, D. J. (1957). Cubic forms over algebraic number fields. *Mathematika*, **4**, 97–101. [9].

Lewis, D. J. (1970). *Systems of diophantine equations.* Symp. Math. IV, INDAM, Rome 1968/1969, 33–43. Academic Press. [G].

Lewis, D. J. (1973). *The distribution of the values of real quadratic forms at integer points.* American Mathematical Society Proceedings of Symposia in Pure Mathematics, **24**, 159–74. [G].

Linnik, Ju. V. (1942, 1943*a*). On the representation of large numbers as sums of seven cubes. *Doklady Akad. Nauk SSSR*, **35**, 162 and *Mat. Sbornik*, **12**, 218–24. [1].

Linnik, Ju. V. (1943*b*). An elementary solution of the problem of Waring by Schnirel'man's method. *Mat. Sb.*, **12**, 225–30. [1].

Linnik, Ju. V. (1945). On the possibility of a unique method in certain problems of "additive" and "distributive" prime number theory. *Dokl. Akad. Nauk SSSR*, **48**, 3–7. [3].

Linnik, Ju. V. (1946). A new proof of the Goldbach–Vinogradov theorem. *Mat. Sb.*, **19**(61), 3–8. [3].

Linnik, Ju. V. (1951). Prime numbers and powers of two. *Trudy Mat. Inst. Steklov*, **38**, 152–169. [G].

Linnik, Ju. V. (1951, 1952). Some conditional theorems concerning binary problems with prime numbers. *Doklady Akad. Nauk SSSR*, **77**, 15–18 and *Izv. Akad. Nauk SSSR Ser. Mat.*, **16**, 503–20. [3]

Linnik, Ju. V. (1953). Addition of prime numbers with powers of one and the same number. *Mat. Sb.*, **32**(74), 3–60. [G].

Liu, M. -C. (1974). Simultaneous approximation of two additive forms. *Proc. Camb. Philos. Soc.*, **75**, 77–82. [G].

Liu, M. -C. (1977). Diophantine approximation involving primes. *J. Reine Angew. Math.*, **289**, 199–208. [G].

Liu, M. -C. (1978). Approximation by a sum of polynomials involving primes. *J. Math. Soc. Jpn.*, **30**, 395–412. [G].

Liu, M. -C. (1979). Approximation by a sum of polynomials of different degrees involving primes. *J. Aust. Math. Soc.*, **27A**, 454–66. [G].

Liu, M. -C., Ng. S. -M. & Tsang, K. -M. (1980). An improved estimate for certain diophantine inequalities. *Proc. Am. Math. Soc.*, **78**, 457–63. [G].

Lloyd, D. P. (1975). Bounds for solutions of Diophantine equations. Ph.D. thesis. University of Adelaide. [G].

Lu, M. -G. & Chen, W. -D. (1965). On the solution of systems of linear equations with prime variables. *Acta Math. Sinica*, **15**, 731–48, translated in *Chin. Math.-Acta*, **7**, 461–79. [G].

Lucke, B. (1926). Zur Hardy–Littlewoodschen Behandlung des Goldbachschen Problems. Dissertation. Math.-naturwiss. Göttingen. [3].

Lursmanashvili, A. P. (1966). Representation of natural numbers by sums of prime numbers. *Thbilis. Sahelmc. Univ. Shrom. Mekh.-Math. Mecn. Ser.*, **117**, 63–76. [3].

Mahler, K. (1957). On the fractional parts of the powers of a rational number II. *Mathematika*, **4**, 122–4. [1].

Mahler, K. (1968). An unsolved problem on the powers of 3/2. *J. Aust. Math. Soc.*, **8**, 313–21. [1].

Malyshev, A. V. & Podsypanin, E. V. (1974). Analytic methods in the theory of systems of Diophantine equations and inequalities with a large number of unknowns. *Algebra, Topology, Geometry*, **12**, 5–50. *Akad. Nauk SSSR Vsesojuz. Inst. Nauk i Tehn. Informacii. Moscow.* [S].

Mardzhanishvili, K. K. (1936, 1937). Über die simultane Zerfällung ganzer Zahlen in *m*-te und *n*-te Potenzen. *Dokl. Akad. Nauk SSSR*, **2**, 263–4 and *Izv. Akad. Nauk SSSR, Ser. Mat.*, 609–31. [7].

Mardzhanishvili, K. K. (1939). Sur un système d'equations de Diophante. *Doklady Akad. Nauk SSSR*, **22**, 467–70. [7].

Mardzhanishvili, K. K. (1940). Sur un problème additif de la théorie des nombres. *Izv. Akad. Nauk SSSR*, **4**, 193–214. [7].

Mardzhanishvili, K. K. (1941). Sur la démonstration du théorème de Goldbach–Vinogradoff. *Dokl. Akad. Nauk SSSR*, **30**, 687–9. [3].

Mardzhanishvili, K. K. (1947). On an asymptotic formula of the additive theory of prime numbers. *Soobscheniya Akad. Nauk Gruzin. SSR*, **8**, 597–604. [G].

Mardzhanishvili, K. K. (1949). On some additive problems with prime numbers. *Uspehi Mat. Nauk*, **4**, 183–5. [G].

Mardzhanishvili, K. K. (1950*a*). On a generalization of Waring's problem. *Soobscheniya Akad. Nauk Gruzin. SSR*, **11**, 82–4. [G].

Mardzhanishvili, K. K. (1950*b*). On a system of equations in prime numbers. *Dokl. Akad. Nauk SSSR*, **70**, 381–3. [G].

Mardzhanishvili, K. K. (1950*c*). Investigations on the application of the method of trigonometric sums to additive problems. *Uspehi Mat. Nauk*, **5**, 236–40. [G].

Mardzhanishvili, K. K. (1951*a*). On the simultaneous representation of pairs of numbers by sums of primes and their squares. *Akad. Nauk Gruzin. SSR. Trudy Mat. Inst. Razmaaze*, **18**, 183–208. [G].

Mardzhanishvili, K. K. (1951*b*). On some additive problems of the theory of numbers. *Acta Math. Acad. Sci. Hungar.*, **2**, 223–7. [S].

Mardzhanishvili, K. K. (1953). On some nonlinear systems of equations in integers. *Mat. Sb.*, **33**, (75), 639–75. [7].

Miech, R. J. (1968). On the equation $n = p + x^2$. *Trans. Am. Math. Soc.*, **130**, 494–512. [G].

Mirsky, L. (1958). Additive prime number theory. *Math. Gaz.*, **42**, 7–10. [S].

Mitsui, T. (1960*a,b*). On the Goldbach problem in an algebraic number field I, II. *J. Math. Soc. Jpn.*, **12**, 290–324 and 325–372.

Montgomery, H. L. (1971). A lemma in additive prime number theory. In *Topics in multiplicative number theory. Lecture Notes in Mathematics*, **227**, Chapter 16. Berlin: Springer–Verlag. [3].

Montgomery, H. L. & Vaughan, R. C. (1973). Error terms in additive prime number theory. *Q. J. Math.*, (2), **24**, 207–16. [3].

Montgomery, H. L. & Vaughan, R. C. (1975). The exceptional set in Goldbach's problem. *Acta Arith.*, **27**, 353–70. [3].

Mordell, L. J. (1932). On a sum analogous to a Gauss's sum. *Q. J. Math.*, **3**, 161–7. [7].

Narasimhamurti, V. (1941). On Waring's problem for 8th, 9th and 10th powers. *J. Indian Math. Soc.*, **5**, 122. [6].

Nechaev, V. I. (1949, 1953). The representation of integers by sums of terms of the form $x(x+1)\ldots(x+n-1)/n!$. *Dokl. Akad. Nauk SSSR*, **64**, 159–62 and *Izv. Akad. Nauk SSSR Ser. Mat.*, **17**, 485–98. [G].

Nechaev, V. I. (1951). Waring's problem for polynomials. *Trudy Mat. Inst. Steklov*, **38**, 190–243. [G].

Nechaev, V. I. (1958). Multinomials with small $G(f)$. *Uch. Zap. Moscow. gor. ped. in-ta*, **71**, 291–300. [G].

Nechaev, V. I. & Telesin, Ju. Z. (1962). On the exact value of $G(f,a)$ for sums of multinomials of the second degree. *Uch. Zap. Moscow. gor. ped in-ta*, **188**, 131–8. [G].

Newman, D. J. (1960). A simplified proof of Waring's conjecture. *Michigan Math. J.*, **7**, 291–5. [1].

Niven, I. (1944). An unsolved case of the Waring problem. *Am. J. Math.*, **66**, 137–43. [1].

Norton, K. K. (1966). On homogeneous diagonal congruences of odd degree. Ph. D. thesis. University of Illinois. [9].

Padhy, B. (1936). Pillai's exact formula for the number $g(n)$ in Waring's problem. *Proc. Indian Acad. Sci.*, **3A**, 341–5. [1].

Pan, C. -T. (1959). Some new results in the additive prime number theory. *Acta Math. Sinica*, **9**, 315–29. [3].

Page, A. (1934*a,b*). On the representation of a number as a sum of squares and products I, III. *Proc. Lond. Math. Soc.*, (2), **36**, 241–56 and **37**, 1–16. [G].

Pillai, S. S. (1936*a,b,c,d*, 1937*a,b*, 1938*a,b,c*). On Waring's problem; I. *J. Indian Math. Soc.*, **2**, 16–44, 131: II. *J. Annamalai Univ.*, **5**, 145–66: III. *Ibid.*, **6**, 50–3: IV. *Ibid.*, **6**, 54–64: V. *J. Indian Math. Soc.*, **2**, 213–14: VI. *J. Annamalai Univ.*, **6**, 171–197 : VII. *Proc. Indian Acad. Sci.*, **9A**, 29–34 : VIII. *J. Indian Math. Soc.*, **3**, 205–20: IX. *Ibid.*, 221–5. [1].

Pillai, S. S. (1940). On Waring's problem $g(6)=73$. *Proc. Indian Acad. Sci.*, **12A**, 30–40. [1].

Pil'tai, G. Z. (1972). On the size of the difference between consecutive primes. *Issled. teor. chisel*, 73–9. [G].

Pitman, J. (1968). Cubic inequalities. *J. Lond. Math. Soc.*, **43**, 119–26. [11].

Pitman, J. (1971*a*), Bounds for the solutions of diagonal inequalities. *Acta Arith.*, **18**, 179–90. [11].

Pitman, J. (1971*b*). Bounds for solutions of diagonal equations. *Acta Arith.*, **19**, 223–47. [9, 11].

Pitman, J. & Ridout, D. (1967). Diagonal cubic equations and inequalities. *Proc. R. Soc. Lond.*, **297A**, 476–502. [11].

Pleasants, P. A. B. (1966*a*). The representation of primes by cubic polynomials. *Acta Arith.*, **12**, 23–45. [G].

Pleasants, P. A. B. (1966*b*). The representation of primes by quadratic and cubic polynomials. *Acta Arith.*, **12**, 131–63. [G].

Pleasants, P. A. B. (1967). The representation of integers by cubic forms. *Proc. Lond. Math. Soc.*, (3), **17**, 553–76. [G].

Prachar, K. (1953*a,b*). Über ein Problem vom Waring–Goldbach'schen Typ. I, II. *Monatsh. Math.*, **57**, 66–74; 113–16. [G].

Prachar, K. (1957). *Primzahlverteilung.* Berlin: Springer–Verlag. [3].

Rademacher, H. (1924*a*). Über eine Erweiterung des Goldbachshen Problems. *Math. Z.*, **25**, 627–57. [3].

Rademacher, H. (1924*b*). Zur additiven Primzahltheorie algebraischer Zahlkörper, I Über die Darstellung totalpositiver Zahlen als Summe von totalpositiven Primzahlen im reell-quadratischen Zahlkörper. *Abh. Math. Sem. Hansischen Univ.*, **3**, 109–63. [G].

Rademacher, H. (1924*c*). Zur additiven Primzahltheorie algebraischer Zahlkörper, II Über die Darstellung von Körperzahlen als Summe von Primzahlen im imaginärquadratischen Zahlkörper. *Abh. Math. Sem. Hansischen Univ.*, **3**, 331–78. [G].

Rademacher, H. (1926). Zur additiven Primzahltheorie algebraischer Zahlkörper, III Über die Darstellung totalpositiver Zahlen als Summen von totalpositiven Primzahlen in einem beliebigen Zahlkörper. *Math. Z.*, **27**, 321–426. [G].

Rademacher, H. (1942). Trends in research: the analytic number theory. *Bull. Am. Math. Soc.*, **48**, 379–401. [S].

Rademacher, H. (1950). Additive algebraic number theory. *Proc. Intern. Congr. Math.*, **1**, 356–62. [S].

Raghavan, S. (1974). On a Diophantine inequality for forms of additive type. *Acta Arith.*, **24**, 499–506. [11].

Ramachandra, K. (1973). On the sums $\sum_{j=1}^{k} \lambda_j f(p_j)$. *J. Reine Angew. Math.*, **262/263**, 158–65. [11].

Ramanujan, C. P. (1963). Cubic forms over algebraic number fields. *Proc. Camb. Philos. Soc.*, **59**, 683–705. [G].

Richert, H. -E. (1953). Aus der additiven Primzahltheorie. *J. Reine Angew. Math.*, **191**, 179–98. [3].

Ridout, D. (1958). Indefinite quadratic forms. *Mathematika*, **5**, 122–4. [11].

Rieger, G. R. (1953*a*). Über eine Verallgemeinerung des Waringschen Problems. *Math. Z.*, **58**, 281–3. [1].

Rieger, G. R. (1953*b,c*). Zur Hilbertschen Lösung des Waringschen Problems: Abschätzung von $g(n)$. *Mitt. Math. Sem. Giessen*, **44**, 1–35. and *Arch. Math.*, **4**, 275–81. [1].

Rieger, G. R. (1954). Zu Linniks Lösung des Waringschen Problems: Abschatzung von $g(n)$. *Math. Z.*, **60**, 213–34. [1].

Roth, K. F. (1949). Proof that almost all positive integers are sums of a square, a positive cube and a fourth power. *J. Lond. Math. Soc.*, **24**, 4–13. [8].

Roth, K. F. (1951). On Waring's problem for cubes. *Proc. Lond. Math. Soc.*, (2) **53**, 268–79. [G].

Roth, K. F. (1951). A problem in additive number theory. *Proc. Lond. Math. Soc.*, (2), **53**, 381–95. [8].

Roth, K. F. (1952). Sur quelques ensembles d'entiers. *C. R. Acad. Sci. Paris*, **234**, 388–90. [10].

Roth, K. F. (1953, 1954). On certain sets of integers I, II. *J. Lond. Math. Soc.*, **28**, 104–9 and **29**, 20–6. [10].

Roth, K. F. (1967*a,b*, 1970, 1972). Irregularities of sequences relative to arithmetic progressions I, II, III, IV. *Math. Ann.*, **169**, 1–25; *ibid.*, **174**, 41–52, *J. Number Theor*, **2**, 125–42; *Periodica Math. Hungar.*, **2**, 301–26. [10].

Rubugunday, R. K. (1942). On $g(k)$ in Waring's problem. *J. Indian Math. Soc.*, **6**, 192–8. [1].

Ryavec, C. (1969). Cubic forms over algebraic number fields. *Proc. Camb. Philos. Soc.*, **66**, 323–33. [G].

Salem, R. & Spencer, D. C. (1942). On sets of integers which contain no three terms in arithmetical progression. *Proc. Natn. Acad. Sci. U.S.A.*, **28**, 561–3. [10].

Salem, R. & Spencer, D. C. (1950). On sets which do not contain a given number of terms in arithmetical progression. *Niew. Arch. Wisk.*, (2), **23**, 133–43. [10].

Sambasiva Rao, K. (1941). On Waring's problem for smaller powers. *J. Indian Math. Soc.*, **5**, 117–21. [6].

Sárközy, A. (1978*a,b,c*). On difference sets of integers I, III, II. *Acta Math. Acad. Sci. Hungar.*, **31**, 125–49; *ibid.*, 355–86; *Ann. Univ. Sci. Budapest Rolando Eötvös, Sect. Math.*, **21**, 45–53. [10].

Sastry, S. & Singh, R. (1955/6). A problem in additive number theory. *J. Sci. Res. Banaras Hindu Univ.*, **6**, 251–65. [8].

Schmidt, E. (1913). Zum Hilbertschen Beweis des Waringschen Theorems. *Math. Ann.*, **74**, 271–4. [1].

Schmidt, W. M. (1976). *Equations over finite fields. An elementary approach. Lecture Notes in Mathematics*, **536**, Berlin: Springer–Verlag. [B].

Schmidt, W. M. (1979*a,b*). Small zeros of additive forms in many variables I, II. *Trans. Amer. Math. Soc.*, **248**, 121–33; *Acta Math.*, **143**, 219–32. [9].

Schmidt, W. M. (1980). Diophantine inequalities for forms of odd degree. *Advances in Math.*, **38**, 128–51.

Schwarz, W. (1960/1, 1961). Zur Darstellung von Zahlen durch Summen von Primzahlpotenzen I, II. *J. Reine Angew. Math.*, **205**, 21–47; **206**, 78–112. [G].

Schwarz, W. (1963). Uber die Lösbarkeit gewisser Ungleichungen durch Primzahlen. *J. Reine Angew. Math.*, **212**, 150–7. [8].

Scourfield, E. J. (1960). A generalization of Waring's problem. *J. Lond. Math. Soc.*, **35**, 98–116. [5,8].

Siegel, C. L. (1944). Generalization of Waring's problem to algebraic number fields. *Am. J. Math.*, **66**, 122–36. [G].

Siegel, C. L. (1945). Sums of *m*th powers of algebraic integers. *Ann. Math.*, (2), **46**, 313–39. [G].

Sinnadurai, J. St.-C. L. (1965). Representation of integers as sums of six cubes and one square. *Q. J. Math.*, (2), **16**, 289–96. [8].

Stanley, G. K. (1929). On the representation of a number as a sum of squares and primes. *Proc. Lond. Math. Soc.*, (2), **29**, 122–44. [G].

Stanley, G. K. (1930). The representation of a number as the sum of one square and a number of *k*-th powers. *Proc. Lond. Math. Soc.*, (2), **31**, 512–53. [G].

Statulevicius, V. (1955). On the representation of odd numbers as the sum of three almost equal prime numbers. *Vilniaus Valst. Univ. Mokslo Darbai Mat. Fiz.-Chem. Mokslu Ser.*, **3**, 5–23. [3].

Stemmler, R. M. (1964). The ideal Waring theorem for exponents 401–200 000. *Math. Comp.*, **18**, 144–6. [1].

Stridsberg, E. (1912). Sur la démonstration de M. Hilbert du théorème de Waring. *Math. Ann.*, **72**, 145–52. [1].

Subhankulov, M. A. (1960). Additive properties of certain sequences of numbers. *Issled. po mat. anal. mech. Uzb.*, 220–41. [G].

Szekeres, G. (1978). Major arcs in the four cubes problem. *J. Aust. Math. Soc.*, **25A**, 423–37. [G].

Szemerédi, E. (1969). On sets of integers containing no four elements in arithmetic progression. *Acta Math. Acad. Sci. Hungar.*, **20**, 89–104. [10].

Szemerédi, E. (1975). On sets of integers containing no k elements in arithmetic progression. *Acta Arith.*, **27**, 199–245. [10].

Tartakovsky, W. (1935). Über asymptotische Gesetze der allgemeinen Diophantischen Analyse mit vielen Unbekannten. *Bull. Acad. Sci. URSS*, 483–524. [9].

Tartakovsky, W. (1958*a,b*). The number of representations of large numbers by a form of "general type" with many variables I, II. *Vestnik Leningrad Univ.*, **13**, 131–54; **14**, 5–17. [9].

Tatuzawa, T. (1955). Additive prime number theory in an algebraic number field. *J. Math. Soc. Jpn.*, **7**, 409–23. [G].

Tatuzawa, T. (1958). On the Waring problem in an algebraic number field. *J. Math. Soc. Jpn.*, **10**, 322–41. [G].

Tatuzawa, T. (1973). On Waring's problem in algebraic number fields. *Acta Arith.*, **24**, 37–60. [G].

Telesin, Yu. Z. (1958). Waring's problem for polynomials of degree 7, 8, 9, 10. *Uch. zap. Moscow. gor. ped. in-ta*, **71**, 301–11. [G].

Thanigasalam, K. (1966). A generalization of Waring's problem for prime powers. *Proc. Lond. Math. Soc.*, (3), **16**, 193–212. [G].

Thanigasalam, K. (1967). Asymptotic formula in a generalized Waring's problem. *Proc. Camb. Philos. Soc.*, **63**, 87–98. [8].

Thanigasalam, K. (1967/1968). On additive number theory. *Acta Arith.*, **13**, 237–58. [G].

Thanigasalam, K. (1969). Note on the representation of integers as sums of certain powers. *Proc. Camb. Philos. Soc.*, **65**, 445–6. [8].

Thomas, H. E. Jr. (1974). Waring's problem for twenty two biquadrates. *Trans. Am. Math. Soc.*, **193**, 427–30. [1].

Tietäväinen, A. (1964). On the non-trivial solvability of some systems of equations in finite fields. *Ann. Univ. Turku. Ser. A.* I, No. 71. [9].

Tietäväinen, A. (1965). On the non-trivial solvability of some equations and systems of equations in finite fields. *Ann. Acad. Sci. Fenn. Ser. A.* I, No. 360. [9].

Tietäväinen, A. (1971). On a problem of Chowla and Shimura, *J. Number Theor.*, **3**, 247–52. [9].

Toliver, R. H. (1975). Bounds for solutions of two simultaneous additive equations of odd degree. Ph.D. thesis. University of Michigan. Ann. Arbor. [G].

Tong, K. -C. (1957). On Waring's problem. *Adv. Math.*, **3**, 602–7. [5].

Trost, E. (1958). Eine Bemerkung zum Waringschen Problem. *Elem. Math.*, **13**, 73–5. [1].

Uchiyama, S. (1961). Three primes in arithmetical progression. *Proc. Jpn. Acad.*, **37**, 329–30. [3].

Vaughan, R. C. (1970). On the representation of numbers as sums of powers of natural numbers. *Proc. Lond. Math. Soc.*, (3), **21**, 160–80. [8].

Vaughan, R. C. (1971). On sums of mixed powers. *J. Lond. Math. Soc.*, (2), **3**, 677–88. [6].

Vaughan, R. C. (1972). On Goldbach's problem. *Acta Arith.*, **22**, 21–48. [3].

Vaughan, R. C. (1973). A new estimate for the exceptional set in Goldbach's problem. *Am. Math. Soc. Proc. Symp. Pure Math.*, **24**, 315–20. [3].

Vaughan, R. C. (1973/1974). A survey of recent work in additive prime number theory. *Sem. Théor. Nombres*, **19**, 1–7. Bordeaux. [S].

Vaughan, R. C. (1974*a*, *b*). Diophantine approximation by prime numbers I, II. *Proc. Lond. Math. Soc.*, (3), **28**, 373–84; 385–401. [11].

Vaughan, R. C. (1975). Mean value theorems in prime number theory. *J. Lond. Math. Soc.*, (2), **10**, 153–62. [3].

Vaughan, R. C. (1977*a*). On pairs of additive cubic equations. *Proc. Lond. Math. Soc.*, (3), **34**, 354–64. [G].

Vaughan, R. C. (1977*b*). Homogeneous additive equations and Waring's problem. *Acta Arith.*, **33**, 231–53. [5, 6, 9].

Vaughan, R. C. (1977*c*). Sommes trigonométriques sur les nombres premiers. *C. R. Acad. Sci. Paris, Sér. A*, **258**, 981–3. [3].

Vaughan, R. C. (1979). A survey of some important problems in additive number theory. *Soc. Math. de France. Astérisque*, **61**, 213–22. [S].

Vaughan, R. C. (1980*a*). A ternary additive problem. *Proc. Lond. Math. Soc.*, **41**, 516–32. [8].

Vaughan, R. C. (1980*b*). *Recent work in additive prime number theory*. Proceedings of the International Congress of Mathematicians, Helsinki, 1978, 389–94. [3].

Veidinger, L. (1958). On the distribution of the solutions of diophantine equations with many unknowns. *Acta Arith.*, **5**, 15–24. [G].

Verdenius, W. (1949). On problems analogous to those of Goldbach and Waring. *Ned. Akad. Wet.*, **52** = *Indag. Math.*, **11**, 255–63. [G].

Vinogradov, A. I. (1955). On some new theorems of the additive theory of numbers. *Dokl. Akad. Nauk SSSR*, **102**, 875–76. [G].

Vinogradov, A. I. (1956). On an almost binary problem. *Izv. Akad. Nauk SSSR, Ser. Mat.*, **20**, 713–50. [G].

Vinogradov, A. I. (1963). On a problem of L. K. Hua. *Dokl. Akad. Nauk SSSR*, **151**, 255–7. [3].

Vinogradov, I. M. (1928*a*). Sur le théorème de Waring. *C. R. Acad. Sci. URSS*, 393–400. [1].

Vinogradov, I. M. (1928*b*). Sur la représentation d'un nombre entier par un polynom à plusiers variables. *C. R. Acad. Sci. URSS*, (7), **1**, 401–14. [1].

Vinogradov, I. M. (1934*a*). A new solution of Waring's problem. *C. R. Acad. Sci. URSS*, (2), **2**, 337–41. [5].

Vinogradov, I. M. (1934*b*). On the upper bound *G*(*n*) in Waring's problem. *C. R. Acad. Sci. URSS*, 1455–69. [5].

Vinogradov, I. M. (1935*a*). Une nouvelle variante de la démonstration du théorème de Waring. *C. R. Acad. Sci. Paris*. **200**, 182–4. [5].

Vinogradov, I. M. (1935*b*). On Waring's problem. *Ann. Math.*, **36**, 395–405. [5].

Vinogradov, I. M. (1935*c*). A new variant of Waring's theory. *Trav. Inst. Steklov*, **9**, 5–15. [5].

Vinogradov, I. M. (1935*d*). On Weyl's sums, *Rec. Math.*, **42**, 521–30. [5].

Vinogradov, I. M. (1935*e*). An asymptotic formula for the number of representations in Waring's problem. *Rec. Math.*, **42**, 531–4. [5].

Vinogradov, I. M. (1937*a*). Representation of an odd number as a sum of three primes. *C. R. Acad. Sci. URSS*, **15**, 6–7. [3].

Vinogradov, I. M. (1937*b*). Some theorems concerning the theory of primes. *Rec. Math.*, **2**, (44), 2, 179–95. [3].

Vinogradov, I. M. (1937*c*). Some new problems of the theory of primes. *C. R. Acad. Sci. URSS*, **16**, 131–2. [G].

Vinogradov, I. M. (1937*d*). A new method in analytic number theory. *Trav. Inst. Steklov*, **10**, 1–122. [5].

Vinogradov, I. M. (1947). The method of trigonometrical sums in the theory of numbers. *Trav. Inst. Steklov*, **23**, translated from the Russian, revised and annotated by Davenport, A & Roth, K. F. (1954). New York: Interscience. [E].

Vinogradov, I. M. (1954). *Elements of number theory*, New York: Dover. Translated from the fifth Russian edition of 1949 by S. Kravetz. [B].

Vinogradov, I. M. (1959). On an upper bound for *G*(*n*). *Izv. Akad. Nauk SSSR*, **23**, 637–42. [7].

Waerden, B. L. van der. (1927). Beweis einer Baudetschen Vermutung. *Niew Arch. Wisk.*, **15**, 212–16. [10].

Walfisz, A. (1941*a*,*b*). Zur additiven Zahlentheorie VII(1), (2). *Soobschenia Akad. Nauk Gruzinskoi SSR*, **2**, 7–14; 221–6. [3].

Watson, G. L. (1951). A proof of the seven cube theorem. *J. Lond. Math. Soc.*, **26**, 153–6. [1].

Watson, G. L. (1953). On indefinite quadratic forms in five variables. *Proc. Lond. Math. Soc.*, (3), **3**, 170–81. [11].

Watson, G. L. (1969). A cubic Diophantine equation. *J. Lond. Math. Soc.*, (2), **1**, 163–73. [G].

Weyl, H. (1916). Über die Gleichverteilung von Zahlen mod Eins. *Math. Ann.* **77**, 313–52. [2].

Whiteman, A. L. (1940). Additive prime number theory in real quadratic fields. *Duke Math. J.*, **7**, 208–32. [G].

Wilson, R. J. (1969). The large sieve in algebraic number fields. *Mathematika*, **16**, 189–204. [5].

Wright, E. M. (1933*a*,*b*). The representation of a number as a sum of five or more squares I, II. *Q. J. Math.*, **4**, 37–51; 228–32. [G].

Wright, E. M. (1934). Proportionality conditions in Waring's problem. *Math. Z.*, **38**, 730–46. [G].

Zuckerman, H. S. (1936). New results for the number *g*(*n*) in Waring's problem. *Am. J. Math.*, **58**, 545–52. [1].

Zulauf, A. (1952*a*). Beweis einer Erweiterung des Satzes von Goldbach–Vinogradov. *J. Reine Angew. Math.*, **190**, 169–98. [3].

Zulauf, A. (1952*b*). Zur additiven Zerfallung natürlicher Zahlen in Primzahlen und Quadrate. *Arch. Math.*, **3**, 327–33. [G].

Zulauf, A. (1953*a*). Über den dritten Hardy–Littlewoodschen Satz zur Goldbachschen Vermutung. *J. Reine Angew. Math.*, **192**, 117–28. [3].

Zulauf, A. (1953*b*, 1954*a*,*b*). Über die Darstellung natürlicher Zahlen als Summen von Primzahlen aus gegebenen Restklassen und Quadraten mit gegebenen Koeffizienten. I, Resultate für genügend gross Zahlen; II, Die Singulare Reihe; III Resultate für "fast alle" Zahlen. *J. Reine Angew. Math.*, **192**, 210–29; **193**, 39–53; **193**, 54–64. [G].

Zulauf, A. (1961). On the number of representations of an integer as a sum of primes belonging to given arithmetical progressions. *Compos. Mat.*, **15**, 64–9. [3].

Index

DATE DUE

GAYLORD			PRINTED IN U.S.A.